About Island Press

Island Press is the only nonprofit organization in the United States whose principal purpose is the publication of books on environmental issues and natural resource management. We provide solutions-oriented information to professionals, public officials, business and community leaders, and concerned citizens who are shaping responses to environmental problems.

In 2007, Island Press celebrates its twenty-second anniversary as the leading provider of timely and practical books that take a multidisciplinary approach to critical environmental concerns. Our growing list of titles reflects our commitment to bringing the best of an expanding body of literature to the environmental community throughout North America and the world.

Support for Island Press is provided by the Agua Fund, The Geraldine R. Dodge Foundation, Doris Duke Charitable Foundation, The William and Flora Hewlett Foundation, Kendeda Sustainability Fund of the Tides Foundation, Forrest C. Lattner Foundation, The Henry Luce Foundation, The John D. and Catherine T. MacArthur Foundation, The Marisla Foundation, The Andrew W. Mellon Foundation, Gordon and Betty Moore Foundation, The Curtis and Edith Munson Foundation, Oak Foundation, The Overbrook Foundation, The David and Lucile Packard Foundation, The Winslow Foundation, and other generous donors.

The opinions expressed in this book are those of the author(s) and do not necessarily reflect the views of these foundations.

Parks and
Carrying Capacity

Parks and Carrying Capacity

Commons Without Tragedy

by
Robert E. Manning

ISLANDPRESS

Washington • Covelo • London

Island Press is a trademark of The Center for Resource Economics.

Earlier versions of several chapters appeared in *Studies in Outdoor Recreation: Search and Research for Satisfaction*, by R. Manning (Corvallis: Oregon State University Press, 1999): chapter 2 is revised and updated from chapter 4; portions of chapters 3 and 5 are abstracted and revised from chapter 5; and portions of chapters 17 and 18 are revised and updated from chapter 12. Reprinted with the permission of Oregon State University Press.

Portions of chapter 6 are revised and expanded from "Use of visual research methods to measure standards of quality for parks and outdoor recreation," *Journal of Leisure Research* 36 (2004): 552–79. Portions of chapter 10 are abstracted and revised from "Estimating day use social carrying capacity in Yosemite National Park," *Leisure: The Journal of the Canadian Association for Leisure Studies* 27 (2003) 77–102. Portions of chapter 11 are abstracted and revised from two articles in *Journal of Park and Recreation Administration* 21 (2003): "Research to guide management of wilderness camping at Isle Royale National Park: Part I–Descriptive Research," 22–42, and "Research to guide management of wilderness camping at Isle Royale National Park: Part II–Prescriptive Research," 43–56. Portions of Chapter 12 are abstracted and revised from "Research to support management of visitor carrying capacity at Boston Harbor Islands," 12 *Northeastern Naturalist* (2005), 201–20. Portions of Chapter 13 are abstracted and revised from "Research to estimate and manage carrying capacity of a tourist attraction: A study of Alcatraz Island," *Journal of Sustainable Tourism* (2002): 388–404.

Appreciation is expressed to the original publishers of this material.

Library of Congress and British CIP Data

Manning, Robert E., 1946–
  Parks and carrying capacity / by Robert E. Manning.
    p. cm.
  ISBN-13: 978-1-55963-104-4 (cloth : alk. paper)
  ISBN-10: 1-55963-104-X (cloth : alk. paper)
  ISBN-13: 978-1-55963-105-1 (pbk. : alk. paper)
  ISBN-10: 1-55963-105-8 (pbk. : alk. paper)
  1. National parks and reserves—Public use. 2. Protected areas—Public use.
  3. National parks and reserves—Management. 4. Protected areas—Management.
  I. Title.
  SB486.P83M36 2007
  333.78'3140973—dc22

                                                    2006035720

Printed on recycled, acid-free paper ✿

Manufactured in the United States of America

10 9 8 7 6 5 4 3 2 1

Contents

PART III. Case Studies of Measuring and Managing Carrying Capacity   99

PART IV. Managing Carrying Capacity   193

PART V. Beyond Parks and Protected Areas   229

Preface

It was the early 1990s. I'm not sure of the date, not even certain of the year. My phone rang and it was a call I hadn't expected, but (as it turns out) I'd been waiting for. The National Park Service (NPS) was ready to get serious about addressing the issue of *carrying capacity*. How much and what kinds of uses and associated impacts could (should) ultimately be accommodated in the national parks? A group of NPS planners was taking the initiative and was looking for support from the academic community.

Since then, we've been working together pretty intensively—planners, managers, and researchers. The first step was development of a conceptual framework to guide carrying capacity analysis and management. Then the framework was tested at Arches National Park, Utah. A program of natural and social science research was conducted to support this application. And we've been carrying on this kind of work ever since, studying the diverse environmental, cultural, and social contexts that comprise the U.S. national park system, adapting and testing research methods, conducting field work, sitting around conference tables interpreting the data we gathered. NPS staff marshaled this work into park plans, while my academic colleagues, graduate students, and I delivered papers at conferences and published in obscure (but important!) scholarly journals.

Now that we've applied this work to more than twenty units of the national park system (representing many more individual sites), it's time to tell this story more coherently. What is carrying capacity, and how can it be defined in an operational way? How can research support measurement and management of carrying capacity? What are the options for managing carrying capacity and how well do they work? Can we develop some case studies that might be used to guide application of carrying capacity in the diverse array of parks and protected areas? Can the approaches developed to address carrying capacity in parks and protected areas be extended to other environmental contexts and issues? This book is intended to help answer these and related questions.

Sabbatical leaves present unusual and welcome opportunities to take on writing projects of this nature, and I am grateful to the University of Vermont for its generous sabbatical program. In particular, I thank Don DeHayes, dean of the Rubenstein School of Environment and Natural Resources, for his long-standing support of my research program. The staff at Golden Gate National Recreation Area provided a welcoming and stimulating place to spend much of my sabbatical year, and I thank Bryan O'Neil, Nancy Hornor, and Mike Savidge. Golden Gate National Parks Conservancy helped provide needed funding, and I am grateful to Greg Moore for this support as well as all the other good work he and his staff are doing.

A number of people inside and outside the NPS have been involved in the development and application of the research, planning, and management described in this book. As noted above, this work was initiated by a group of NPS planners, including Marilyn Hof, Jim Hammett, Gary Johnson, and Michael Rees, and associated scientists, including Dave Lime (University of Minnesota), Jane Belnap (U.S. Geological Survey), and the author. Kerri Cahill now provides capable leadership for this effort in the NPS. Several NPS staff at Arches National Park were instrumental in the initial application of this work, including Noel Poe, Jim Webster, and Karen McKinlay-Jones. Park staff who made significant contributions to other applications include Charlie Jacobi, David Manski, and John Kelly (Acadia National Park); Terri Thomas, Michael Savidge, Nancy Hornor, and Mia Monroe (Golden Gate National Recreation Area, including Alcatraz Island and Muir Woods National Monument); Dave Wood and Bruce Rodgers (Canyonlands National Park); Linda Jalbert (Grand Canyon National Park); Marjorie Smith (Saratoga National Historical Park); John Sacklin (Yellowstone National Park); Diane Dayson, Cynthia Garrett, and Richard Wells (Statue of Liberty National Monument); Gary Johnson (Blue Ridge Parkway); Rita Hennessey (Appalachian National Scenic Trail); Patty Trap (Mesa Verde National Park); Bruce Jacobson and George Price (Boston Harbor Islands National Recreation Area); Russell Galipeau, Jan van Wagtendonk, Jerry Mitchell, and Laurel Boyers (Yosemite National Park); Jeff Troutman and Shannon Skibeness (Kenai Fjords National Park); Ann Mayo Kiely (Isle Royale National Park); Sara Peskin and Charles Markis (Sagamore Hill National Historic Site); Frank Baublits (Haleakala National Park); Jeff Bradybaugh (Zion National Park); Mike Tranel, Joe Van Horn, Phillip Hooge, Tom Meier, and Carol McIntyre (Denali National Park and Preserve); Vicki Stinson (Hawaii Volcanoes National Park); and Nancy Finley (Cape Cod National Seashore). Nora Mitchell of the NPS Conservation Study Institute, Jerrilyn Thompson of the Great Lakes/Northern Forest Cooperative Ecosystem Studies Unit, Darryll Johnson of the Pacific Northwest Cooperative Ecosystem Studies Unit, and Mary Foley of the U.S. National Park Service, Northeast Region, were generous in providing an administrative mechanism for conducting this work.

I have been fortunate to have a number of colleagues at academic institutions, government agencies, and other organizations as I have pursued the work described in this book. These colleagues include Dave Lime (University of Minnesota), Jeff Marion (U.S. Geological Survey), Yu-Fai Leung (North Carolina State University), Alan Graefe (Penn Sate University), Gerard Kyle (Texas A&M University), David Cole (Aldo Leopold Wilderness Research Institute), Bill Stewart (University of Illinois), Martha Lee (Northern Arizona University), Jonathon Taylor (U.S. Geological Survey), Darryll Johnson (U.S. National Park Service), Mark Vande Kamp (University of Washington), Wayne Freimund and Bill Borrie (University of Montana), Ben Minteer and Megha Budruk (Arizona State University), Peter Newman (Colorado State University), Steve Lawson (Virginia Tech University), Daniel Laven (U.S. National Park Service), Mary Watzin (University of Vermont), and Bill Byrne (David Evans and Associates, Inc.).

Staff and graduate students in the Park Studies Laboratory at the University of Vermont were instrumental in conducting this program of research and include William Valliere, Ben Minteer, Steven Lawson, Peter Newman, Megha Budruk, Benjamin Wang, Jennifer Morrissey, Daniel Laven, James Bacon, Jeffrey Hallo, Rebecca Stanfield McCown, Daniel Abbe, and Logan Park. Much of the material presented in this book is the result of our close collaboration, which I value both professionally and personally. The references cited in the book are a clear manifestation of the team approach we take in the lab and the important contributions of everyone who works there.

Barbara Dean at Island Press supported this book project from the beginning and skillfully guided it through the publication process.

Introduction

And another, and another. . . .

By the early 1990s, the number of visits to the U.S. national park system had topped 250 million a year and was continuing its historic upward trend. That so many people were interested in and attracted to the national parks was something to celebrate. But it also presented serious challenges. National parks, of course, are to be protected as well as used, and the impacts caused (intentionally or unintentionally) by visitors, when multiplied by hundreds of millions each year, presented a serious threat to the integrity of the parks. Fragile vegetation was being trampled, soils eroded, water and air polluted, wildlife disturbed, soundscapes disrupted, and cultural resources diminished. In the process, the quality of the visitor experience was being threatened through crowding and congestion, conflicting uses, and the aesthetic consequences of resource degradation.

The issue of how much use can (should) ultimately be accommodated in parks and protected areas is conventionally called *carrying capacity* in the professional literature, and the National Park Service (NPS) resolved in the early 1990s to address this issue. This effort was led by a group of NPS planners and was supported by several government and university scientists. Based on the scientific and professional literature, a framework was devised to analyze and manage carrying capacity in the national parks and related areas. The framework was called *Visitor Experience and Resource Protection* (now commonly referred to by its acronym *VERP*) as a positive expression of its intentions: the framework was designed to identify and protect what is important about parks and not to inherently limit visitor use (though such limits are needed in some places and at some times). VERP defines indicators and standards for park resources and the quality of the visitor experience, establishes procedures for monitoring those conditions, and requires management actions to ensure that standards are maintained.

VERP was initially applied to Arches National Park, Utah, and this application was supported by a program of natural and social science research (Hof et al. 1994; Manning et al. 1996a; Manning et al. 1996b; Belnap 1998; Manning 2001). A carrying capacity plan was developed for this park, the first such plan in the national park system (National Park Service 1995). A handbook for applying the VERP framework was then developed for NPS planners and managers along with supporting materials (National Park Service 1997; Anderson et al. 1998; Lime et al. 2004). The VERP framework is now being incorporated into planning and management for all units of the national park system.

Just as at Arches National Park, applications of VERP at other units of the national park system are being supported by research. Studies have been conducted at more than twenty units of the national park system, encompassing dozens of sites within these parks. These areas reflect the diversity of the park system and range from "crown jewel" parks like Yellowstone, Yosemite, and Grand Canyon to historical and cultural areas such as the Statue of Liberty, Alcatraz Island, and Mesa Verde National Park. This research has adopted, adapted, and applied an array of theory and methods from a host of academic disciplines, including sociology, psychology, ecology, economics, statistics, business management, landscape architecture, and computer science.

This work has contributed to development of carrying capacity plans for several national parks, and a number of papers describing this program of research, planning, and management have been presented at academic and professional conferences and published in scholarly journals. But this work is scattered across the academic and professional landscape. The purpose of this book is to integrate and synthesize this work into a more comprehensive and coherent volume.

The book begins with a historical and conceptual treatment of carrying capacity. Garrett Hardin's 1960s environmental classic, "The Tragedy of the Commons" (from which the book's title is adapted and the epigraphs at the beginning of each chapter are extracted), offers a prescient and powerful entree into part 1 of the book: national parks are a classic manifestation of the challenges associated with managing common property resources and associated carrying capacity–related issues. But concern over carrying capacity began long before that and is a derivation of the most fundamental question in all of conservation: how much can we use the environment without spoiling it? In contemporary terminology, carrying capacity is now morphing into *sustainability* and is expanding into many sectors of environmental management and modern life more broadly. The remainder of part 1 describes the evolution of our understanding of carrying capacity, how it can be defined in an operational way, and development of conceptual frameworks to address it in the context of parks and protected areas.

Part II describes and illustrates a series of research approaches that can be used to help analyze and manage carrying capacity. Application of carrying capacity will always require some element of management judgment, but such judgments should be as informed as possible. The research approaches described in part 2—qualitative and quantitative surveys, normative theory and methods, visual research approaches, tradeoff analysis, simulation modeling—are designed to help inform application of carrying capacity. Portions of the chapters comprising part 2 are (unavoidably) technical, but these research approaches are illustrated in more applied contexts in part 3.

A series of case studies is the focus of part 3. These case studies describe programs of natural and social science research designed to support analysis and management of carrying capacity at eight diverse units of the national park system. The context of each park gives rise to a variety of research adaptations, potential indicators and standards of park resources and experiences, and related monitoring and management issues.

Attempts to address carrying capacity would ring hollow without the ability to manage visitor use of parks and associated resource and social impacts. In fact, there are a number of management alternatives available, and these are outlined and evaluated in part 4. Limitations on visitor use are an important part of the management arsenal, but other less draconian alternatives are possible.

Part 5 extends carrying capacity beyond parks and protected areas and addresses environmental management more broadly. The conceptual foundations of carrying capacity can (and probably should) be applied to a range of environmental issues and areas. In fact, the process of formulating indicators and standards, monitoring, and adaptive management—the foundational components of carrying capacity—are being applied in an increasing number of environmental and natural resources fields to address the growing urgency of sustainability. A case study of environmental management in the Lake Champlain Basin (Vermont, New York, and Quebec) is used as an example.

The book ends with several conclusions about carrying capacity and its application to parks and protected areas and beyond. Following Hardin's challenge that the tragedy of the commons (and its close cousin carrying capacity) can only be resolved through the wisdom and courage of collective social action—"mutual coercion, mutually agreed upon"—we need to manage parks and protected areas deliberately. The conceptual foundations, research approaches, and management practices outlined in this book offer some tools that can help facilitate and inform this process, and the case studies in the national park system suggest models for their application.

PART I

From Commons to Carrying Capacity

Common property resources and carrying capacity are long-standing and foundational issues in environmental management. These issues are closely related and address the most fundamental question in environmental thought: how much can we use the environment without spoiling what we find most valuable about it? The historical lineage of these issues can be traced back through centuries, but their emergence in contemporary environmental literature might best be attributed to Garrett Hardin's paper, "The Tragedy of the Commons," published in *Science* in 1968. Hardin asserted that without deliberate management action, "mutual coercion, mutually agreed upon," human use of common property resources would inevitably exceed carrying capacity, leading to tragic environmental and associated consequences. He and others include parks and protected areas as classic examples of common property resources.

An extensive scientific and professional literature has developed on these and related issues over the past several decades. Interpretation of both the tragedy of the commons and carrying capacity has evolved to suggest that they are less deterministic and more normative than originally envisioned. That is, when applied in human contexts, societal norms and values—expressed in terms of desired environmental and social conditions—can provide a theoretical and empirical basis for analyzing and managing carrying capacity and resolving the tragedy of the commons.

This interpretation is now being applied in a number of environmental fields, including management of parks and protected areas. Carrying capacity frameworks for parks and related areas have been developed to guide this work. Common and important components of these frameworks include (1) development of management objectives (or desired conditions) and associated indicators and standards; (2) monitoring of indicator variables; and (3) management actions designed to maintain standards. A growing body of research and management experience has begun to identify desirable characteristics of indicators and standards and compile examples of indicators and standards that apply to a range of park resources, experiences, and management contexts.

CHAPTER I

The Tragedy of the Commons

Freedom in a commons brings ruin to all.

In 1968, a haunting paper—"The Tragedy of the Commons"—was published in the prestigious journal *Science* (Hardin 1968). Now a foundational piece of the environmental literature, this paper identified a set of environmental problems—issues of the "commons"—that have no technical solutions but must be resolved through social action. Hardin's ultimate prescription for managing the commons was "mutual coercion, mutually agreed upon": without such collective action, environmental (and related social) tragedy is inevitable.

Hardin began his paper with an illustration using perhaps the oldest and simplest example of an environmental commons, a shared pasture:

> Picture a pasture open to all. It is expected that each herdsman will try to keep as many cattle as possible on [this] commons. . . . What is the utility of adding one more animal? . . . Since the herdsman receives all the proceeds from the sale of the additional animal, the positive utility [to the herdsman] is nearly +1. . . . Since, however, the effects of overgrazing are shared by all the herdsmen, the negative utility for any particular . . . herdsman is only a fraction of −1. Adding together the . . . partial utilities, the rational herdsman concludes that the only sensible course for him to pursue is to add another animal to [the] herd. And another; and another. . . . Therein is the tragedy. Each man is locked into a system that [causes] him to increase his herd without limit—in a world that is limited. . . . Freedom in commons brings ruin to all. (1244)

Hardin went on to identify and explore other examples of environmental commons, ultimately addressing human population growth. However, one of his examples of the tragedy of the commons—one that resonates more urgently each year—is national parks and protected areas:

> The National Parks present another instance of the working out of the tragedy of the commons. At present, they are open to all without limit. The parks themselves are limited in extent—there is only one Yosemite Valley—whereas population seems to grow without limit. The values that visitors seek in the parks are steadily eroded. Plainly, we must soon cease to treat the parks as commons or they will be of no value to anyone. (1245)

The tragedy of the commons has become one of the most compelling and powerful ideas in the environmental literature. The original paper has been republished in over one hundred environmental and public policy–related anthologies and has stimulated an enormous body of research and writing. A recent bibliography on papers related to issues of managing environmental and related commons includes over thirty-seven thousand citations (Hess 2004). This work has been applied to a growing list of commons-related resources and issues, including wildlife and fisheries, surface and ground water, range lands, forests, parks, the atmosphere, climate, oil and other energy resources, food, biodiversity, and population. The conceptual foundation of the tragedy of the commons has even been extended to a growing array of public resources that are not necessarily environmentally related, such as education (J. Brown 2000), medicine (R. Lewis 2004), and the infosphere or cyberspace (Greco and Floridi 2004). Recognizing the importance of common property resources and the issues identified by Hardin's 1968 paper, a special issue of *Science* was published in 2003 commemorating the thirty-fifth anniversary of publication of "The Tragedy of the Commons" and assessing the growing scientific and professional literature it has spawned.

Hardin and others have noted that the issue of managing common property resources has a long history. In fact, nascent interest in the "commons" was expressed by Aristotle, who wrote, "What is common to the greatest number gets the least amount of care. Men pay most attention to what is their own: they care less for what is common" (quoted in Hardin and Baden 1977, xi). The first modern expression of the commons issue is generally credited to Lloyd (1833) who published two lectures in England titled *On the Checks to Population*, which suggested the environmental degradation caused by unfettered population growth and the associated inability of the Earth to support very large numbers of humans. More contemporary, scientifically based explications of the commons were first offered in the 1950s in the context of ocean fisheries (Gordon 1954; Scott 1955).

Common property resources can be defined technically as having several characteristics (Ostrom et al. 1999; Feeny et al. 1990; Ostrom and Ostrom 1977; Greco and Floridi 2004). First, as the term suggests, ownership of the resource is held in common, often by a large number of owners who have independent rights to use the resource. Second, control of access to the resource is problematic for several potential reasons, including the large size or area of the resource, its pervasive character, its migratory nature, or its political intransigence. Third, the level of exploitation by one user adversely affects the ability of other users to exploit the resource. Hardin (1968) and others have noted that in addition to conventional common property resources in which tangible (e.g., forage, fish) and intangible (e.g., enjoyment) benefits are extracted from a resource, there are also "reverse" commons in which pollution is deposited into a resource that is owned in common, such as the oceans and the atmosphere. Management problems associated with common property resources typically arise and need attention when demand for access to the resource exceeds its supply.

Mutual Coercion, Mutually Agreed Upon

Beyond describing common property resources and their potentially tragic consequences, Hardin also discussed how this tragedy might be averted. First, he asserted that there are (ultimately) no technological solutions to the tragedy of the commons. Increased efficiency of resource use might postpone the need to address this issue, but some limitations on resource use will eventually be required. (In fact, Hardin suggested that in some cases, such as ocean fisheries, improved technology in the form of more efficient and effective harvesting may exacerbate or hasten the tragedy of the commons.) Hardin suggested that only two forms of ownership or management could address the tragedy of the commons: private or government ownership. Private ownership "internalizes" both the benefits and costs of exploitation (benefits and costs are both borne by the owner) leading to more rational and productive management. Government ownership allows for a broad and long-term management perspective that is focused on the ultimate welfare of society as a whole (as opposed to an individual), thus offering protection for resources that are ultimately important to society.

Western society, and the United States in particular, relies heavily on private ownership and management of resources to guide production and consumption of goods and services. This approach is inherent in the capitalist, free market economic system. According to the concept of "the invisible hand" proposed by Adam Smith, the decisions of individuals in a free market economy lead to outcomes that ultimately benefit society at large (Smith 1776). While this notion is generally accepted as valid, there are notable exceptions in which government (or social) action is required to address a number of "market failures" or "externalities" (Farley and Daly 2004). For example, the full costs of pollution (e.g., wastes from production processes that are emitted into the atmosphere) are sometimes not paid by producers, leading to below-cost pricing and overproduction of such goods and services and resulting pollution levels that may be harmful to society. Moreover, in some cases it may be difficult to exclude potential users from selected goods and services. National defense and parks and protected areas are representative examples. In such cases, private entities cannot capture the full benefits of producing such goods and services, leading to undersupply. In the case of these market failures, social action (usually in the form of government control) is required through regulation of the free market economy (e.g., laws against excessive pollution, taxes on pollution) and direct government production or management (e.g., a national military, a national park system).

These types of social actions are manifestations of "mutual coercion, mutually agreed upon" that Hardin suggests are ultimately needed to resolve the tragedy of the commons. They are limitations on resource use that apply to all potential users and that all (or at least a majority of) users agree are needed. Parking meters to regulate the use of parking spaces in cities (a common property resource) are a simple example of such "mutual coercion, mutually agreed upon." Without such regulatory institutions, the ability to park in most cities would be unpredictable and chaotic. While these "coercions" may be distasteful because they limit individual freedom, they are needed to protect the greater welfare of society. In rationalizing such limitations, Hardin suggests that "Freedom is the recognition of necessity" (which he attributes to the philosopher Hegel). Only by instituting the mechanisms that will ensure our ultimate well-being will we be truly free to pursue our higher aspirations, both as individuals and as a society.

Social Norms

Early theoretical work suggested that the "mutual coercion, mutually agreed upon" required to resolve the tragedy of the commons can (perhaps must) be a natural outgrowth of cultural or social norms. Norms are a long-standing theoretical construct in the social sciences and represent rules (both formal and informal) that guide human behavior. Such norms are needed in society to provide order and predictability to social life. Most people abide by social norms recognizing that they serve a needed purpose and because formal or informal sanctions in the form of rewards and punishments apply to associated behavior. Norms are discussed in more detail in chapter 5.

More recent empirical research has found that many societies have developed social norms to guide management of common property resources. In this way, there are many examples in which common property resources have been managed in accordance with carrying capacity over very long periods of time (Netting 1972; Netting 1976; McKean 1982; Maass and Anderson 1986; Glick 1970; H. Lewis 1980; Ostrom 1990; National Research Council 1986). Many of the examples described in the literature apply to indigenous cultures and subsistence economies, and nearly all are in the context of local communities or small scales. But some are found in the context of contemporary market-based economies as well. For example, fishermen in Maine have developed informal means of regulating the size and allocation of the harvest to help ensure economically and ecologically sustainable levels of lobster populations (Wilson 1977).

The literature on game theory is also suggestive of the role of communication and cooperation as a means of solving the tragedy of the commons and related social problems (Kuhn 2003; Greco and Floridi 2004; Muhsam 1977). The classic case of the "prisoner's dilemma" is illustrative. While there are many variations of this "game," the most elemental version is a scenario in which two suspects in a crime are arrested and held in isolation from one another. Each prisoner has the option of maintaining silence, confessing, or implicating the other, and there are logical rewards and punishments associated with each of these choices. Game theory suggests that if each prisoner acts strictly according to his own self-interest, the rewards to each will be less than could be attained if they acted cooperatively, and this is borne out in a number of empirical experiments. However, if players of the game are allowed to communicate, or if they "learn" by playing the game multiple times, they will begin to cooperate, thereby overcoming their original, limiting self-interest.

Findings from the theoretical and empirical research suggest that the model of human behavior or rationality originally posited by Hardin may be too rigid and limiting, at least in certain contexts. The purely "economic" rationality described by Hardin does not account for more altruistic human tendencies that can lead to constraints on individual behavior in favor of the greater interests of society, nor does it recognize the possibility of "enlightened self-interest" as individuals restrain their short-term actions to achieve longer-term outcomes that are in their own interest, as well as those of society at large. However, this body of research suggests that informal norms may not always be sufficient to manage common property resources and that this may be especially true at scales that transcend local issues and communities. Thus, it has been concluded that:

> Participants or external authorities must deliberately devise . . . rules that limit who can use a [common property resource], specify how much and when that use will be allowed, create and finance formal monitoring arrangements, and establish sanctions for nonconformance. (Ostrom et al. 1999)

Given the hopefulness of potential solutions—both informal and formal— to the problems of managing common property resources, it has been suggested that the tragedy of the commons is shifting to the "drama of the commons" as researchers and managers try to identify and create the conditions that lead to more sustainable use and management of common property resources (Deitz 2005; National Research Council 2002).

From Commons to Carrying Capacity

The tragedy of the commons is based on an assumption, either explicit or implicit, that there are environmental limits to population and related economic growth. More specifically, concern over the tragedy of the commons is based on the assumption that increasing exploitation of resources will lead to unacceptable environmental (and related social) degradation and ultimately undermine the ability of the natural environment to support life, or at least some minimum quality of life. In technical (and increasingly popular) terminology, this assumption is addressed under the rubric of carrying capacity, and the concept of carrying capacity has become one of the most important and long standing ideas in environmental management.

Most discussions of carrying capacity date its "modern" emergence to an essay published by Thomas Malthus in 1798 titled *An Essay on the Principle of Population* (Malthus [1798] 2003). This essay hypothesized that human population tends to grow in an exponential fashion, but that food production is limited to arithmetic growth, as illustrated in figure 1.1. In this way, the supply of food presents an ultimate limit to population growth, and if these limits are not respected, the result will be (in the words of Malthus) substantial human "vice and misery" and related "positive checks." Malthus's ideas about limits to population and economic growth have become foundational concepts of the contem-

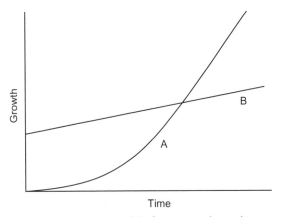

Figure 1.1. Malthus's model of exponential population growth (A) versus arithmetic growth of food resources (B).

porary environmental movement. Popular books such as *The Population Bomb* (Ehrlich 1968), *The Limits to Growth* (Meadows et al. 1972), and *How Many People Can the Earth Support?* (Cohen 1995) are important manifestations of this idea. Indeed, Hardin (1968) references Malthus in his original explication of the tragedy of the commons. Based on this lineage, contemporary environmentalists are sometimes referred to as "neo-Malthusians."

There have also been many academic and scientific treatments of carrying capacity. An early, important paper theorized that population growth (of both humans and other animals) can be characterized by a sigmoid curve defined by the following equation (Pearl and Reed 1920):

$$\frac{dN}{dt} = rN \frac{(K - N)}{K}$$

where N = population size
t = time
r = rate of population growth
K = an asymptote (a tangent to a curve)

The curve derived from this equation is illustrated in figure 1.2. Known as the "logistic growth equation" (and resulting "logistic growth curve"), this mathematical formulation specifies that population grows slowly at first, then faster and faster until it reaches an inflection point associated with approaching environmental limits. After this, population grows more and more slowly as it approximates a horizontal asymptote. This asymptote, often denoted as K, represents carrying capacity and is based on some ultimately limiting factor in the environment (e.g., food, space) (Price 1999; Seidl and Tisdell 1999). First pub-

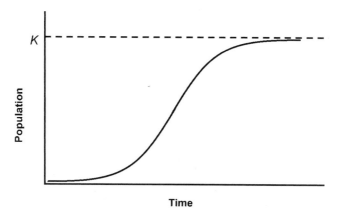

Figure 1.2. Logistic population growth curve.

lished in the modern scientific literature by Pearl and Reed (1920), the authors were unaware at that time that the logistic growth equation had been developed much earlier by Verhulst (1838).

Publication of the paper by Pearl and Reed (1920) sparked immediate interest in, and application of, carrying capacity and its mathematical formulation in the contexts of range and wildlife management and, more generally, in ecology. An initial study applied the concept to grazing of domestic herbivores (Hadwen and Palmer 1922), and this approach was later applied to wildlife in Leopold's (1933) classic text on wildlife management. Following this, Odum (1953) adopted and emphasized the concept of carrying capacity, as denoted by K, in his influential textbook on ecology. However, the body of empirical work on carrying capacity has met with mixed findings (Price 1999). In laboratory experiments with simple life forms, population growth tends to follow the dictates of the logistic growth model as population expansion is controlled by limiting factors such as food and space. However, in "real world" applications with higher life forms, findings are more variable, in that population growth tends to be mediated by a number of factors, including natural variability in environmental conditions and interspecies competition.

From K to I

Just as the complexity of carrying capacity increases as one moves from lower to higher life forms, its complexity is magnified again as it addresses issues of human population and related economic growth. In the context of humans, carrying capacity is now widely recognized as being strongly mediated by a number of social and institutional issues and associated questions (Seidl and Tisdell 1999; Cohen 1995; Cohen 1997; Read and LeBlanc 2003; Davidson 2000; Daily and Ehrlich 1992; Monte-Luna et al. 2004). In this sense, carrying capacity can be most appropriately interpreted as a normative or value-laden concept. For example, to consider the human carrying capacity of an area, one would have to answer questions such as the following: What level of material well-being should be maintained? How should this material well-being be distributed among the population? What level of technology should be applied? What level of environmental protection should be achieved? What social and political institutions should be applied? What time period should be considered? (Cohen 1995; Cohen 1997).

Human carrying capacity is not devoid of natural constraints, but these constraints must be considered in the context of human values and related choices. Thus, carrying capacity as applied to humans is less rigid, positivist, mechanistic, and deterministic than traditional models such as the logistic growth equation might suggest. Recent treatments of human carrying capacity have begun to address this issue by suggesting that it is the impacts of human population and economic growth that need to be emphasized instead of the traditional focus on population numbers per se (Daily and Ehrlch 1992; Seidl and Tisdell 1999). That is, human population and related economic growth has impacts (I) on the environment, and these impacts are what may ultimately dictate maximum (or acceptable) population and economic growth. Moreover, the maximum acceptable levels of such impacts are largely a function of human values as manifested in these types of issues and questions. In this way, carrying capacity analysis and management is evolving from its traditional emphasis on defining maximum population size (K) to defining the conditions under which this population chooses to live (I) (I suggesting the ecological and social impacts of human population growth and related economic development). Under this interpretation, human carrying capacity might be equally well denoted as *social* carrying capacity:

> Actually, the newly titled concept of human or social carrying capacity implies a deep transformation and deviation from the initial biological and demographical positivist concept. (Seidl and Tisdel 1999, 403)

A related way of describing this evolution in thinking is by redefining K. In its original context of the logistic growth model, K represents the environmental limits of population and related economic growth, presumably at some subsistence-related level of existence. However, humans may choose to live at higher levels of material and environmental well-being, and such normative or value-based choices might be symbolized by variations of K. This notion is illustrated in figure 1.3, which uses the symbols Kb and Ks to represent biophysical and social carrying capacity, respectively (Seidl and Tisdell 1999; Ehrlich and Holdren 1971; Hardin 1986; Daily and Ehrlich 1992). Ks represents a conscious choice to stabilize or manage population and related economic growth at a level that is, by definition, lower than that at the margins of ecological limits. Such a choice is presumably related to a desire for some minimum quality of life. Anticipating this train of logic, Hardin (1986, 603), wrote that "carrying capacity is inversely related to the quality of life." This notion has also been used in anthropological studies and is denoted by the symbols K and K^*, where the latter represents the levels at which selected human populations have been found to stabilize at a point that is below that which is ecologically possible (Read and LeBlanc 2003).

Carrying capacity has been subject to a considerable body of theoretical and empirical investigation, and this work has been synthesized in a number of important reviews (Edwards and Fowle 1955; Dasman 1964; Caughley 1979; Dhondt 1988; Price 1999; Seidl and Tisdell 1999; McLeod 1997; Monte-Luna et al. 2004). Many authors conclude that carrying capacity is a vague, contextual, and controversial concept that does not offer the empirical guidance it

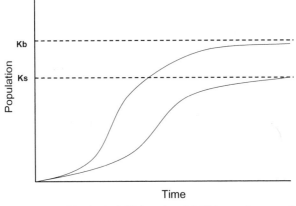

Figure 1.3. Biophysical (K_b) and social (K_s) carrying capacity
(from Seidl and Tisdell, 1999).

seems to suggest, especially in the context of its application to human population and related economic growth. However, recent formulations of carrying
capacity that more explicitly recognize its normative character offer a useful con-
ceptualization. These conceptual models address the factors that affect human
impacts on the environment, judgments about the acceptability of those impacts
and associated environmental and social conditions, and the level and type of
human population and related economic growth that might best be maintained
given desired environmental and related living conditions.

Carrying Capacity of the Commons

The tragedy of the commons and carrying capacity are closely related concepts
that have long and evolutionary histories in environmental management. The
tragedy of the commons is built upon an assumption that there are limits to
human growth and exploitation of the environment (i.e., carrying capacity), and
that carrying capacity must be analyzed and managed to resolve the tragedy of
the commons. Both concepts have evolved from rather rigid, deterministic
frameworks to more normative, contextual notions. In addition to the ecological constraints that might ultimately apply to human population and economic
growth, there are important issues of societal values that must be addressed in
analyzing and managing carrying capacity and resolving the tragedy of the commons. Societal norms and values provide a theoretical and empirical foundation
for defining the environmental and related social conditions upon which carrying capacity must be determined and the management actions—mutual coer-

cion, mutually agreed upon—needed to avert the tragedy of the commons. This contemporary approach to carrying capacity and the commons is being applied in a number of professional fields, including management of parks and protected areas, and this work is described in the following chapter.

CHAPTER 2

Carrying Capacity of Parks
and Protected Areas

The difficulty of defining the optimum is enormous.

Expanding use of national parks and related areas and the growing popularity of outdoor recreation more generally has created a tradition of concern about appropriate use levels of parks, forests, lakes, and other outdoor recreation areas. As noted in chapter 1, these areas fall under the definition of common property resources. Most parks and related areas have been established for public use and appreciation. However, they must also be protected. The two-fold mission of the U.S. national parks stated in the National Park Service Organic Act of 1916 offers the classic expression of this inherent tension as it mandates that national parks are to be managed

> to conserve the scenery and the natural historic objects and the wildlife therein and to provide for the enjoyment of the same in such manner and by such means as will leave them unimpaired for the enjoyment of future generations.

The number of visits to U.S. national parks, as shown in figure 2.1, are indicative of trends in use of parks and protected areas. While use in any one year is subject to ebbs and flows in response to political and economic events, the overall upward trend in use over many decades is obvious. But how much use can ultimately be accommodated in national parks and related areas? What is the *carrying capacity* of parks and outdoor recreation areas?

As suggested in the previous chapter, the fundamental issues underlying the concept of carrying capacity have a long history in human and environmental

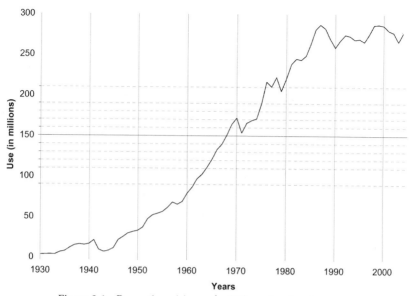

Figure 2.1. Recreation visits to the U.S. national park system.

affairs. In particular, carrying capacity has an especially rich tradition in several of the natural resources professions that substantially predates its application to parks and protected areas. The term has received wide use in wildlife and range management where it can be generally defined as the number of animals of any one species that can be maintained in a given habitat (Dasmann 1964). The principle of carrying capacity also is reflected in the historical concept of *sustained yield* in forestry.

Perhaps the first suggestion for applying the concept of carrying capacity to parks and related areas was recorded in the mid-1930s. A National Park Service (NPS) report on policy recommendations for parks in the California Sierras posed the question, "How large a crowd can be turned loose in a wilderness without destroying its essential qualities?" (Sumner 1936). Later in the report, it was suggested that recreational use of wilderness be kept "within the carrying capacity." A decade later, a paper on recreational use of forest lands suggested that "in all forest recreation, but particularly in zones of concentrated use, carrying capacity is important" (J. V. Wagar 1946). With obvious links to the more traditional use of carrying capacity in wildlife biology, the paper went on to note that "we suspect that humans have saturation points akin to those shown by pheasant and quail." A follow-up article listed carrying capacity as one of eight major principles in recreation land use:

> Forestry, range management, and wildlife management are all based
> upon techniques for determining optimum use and limiting harvest

> beyond this point. . . . Recreation belongs in the same category. (J.
> V. Wagar 1951, 433)

The concept of carrying capacity became a more formal part of the parks and outdoor recreation field as a result of its prominence in the deliberations and writings of the Outdoor Recreation Resources Review Commission (1962), the first national, comprehensive review of the field.

The Science of Carrying Capacity

The first rigorous scientific application of carrying capacity to parks and related areas came in the early 1960s with a conceptual monograph (J. A. Wagar 1964) and a preliminary empirical treatment (Lucas 1964). Perhaps the major contribution of Wagar's conceptual analysis was the expansion of carrying capacity from its dominant emphasis on environmental concerns to a dual focus including social or experiential considerations:

> The study reported here was initiated with the view that the carrying capacity of recreation lands could be determined primarily in terms of ecology and the deterioration of areas. However, it soon became obvious that the resource-oriented point of view must be augmented by consideration of human values. (J. A. Wagar 1964, preface)

Wagar's point was that as more people visit a park or related outdoor recreation area, not only are the environmental resources of the area affected, but also the quality of the recreation experience. Thus, carrying capacity was expanded to include consideration of the social environment as well as the ecological environment. The effects of increasing use on recreation quality were illustrated by Wagar by means of hypothetical relationships between increasing use level and visitor satisfaction.

A preliminary attempt to estimate the recreation carrying capacity of the Boundary Waters Canoe Area, Minnesota, followed shortly, and researchers found that perceptions of crowding varied by different user groups (Lucas 1964). Paddling canoeists were found to be more sensitive to crowding than motor canoeists, who were, in turn, more sensitive to crowding than other motor boaters. A range of carrying capacities was estimated depending upon these different relationships.

Wagar's original conceptual analysis hinted at a third element of carrying capacity, and this was described more explicitly in a subsequent paper (J. A. Wagar 1968). Noting a number of misconceptions about carrying capacity, it was suggested that carrying capacity might vary according to the amount and type of management. For example, the durability of park resources might be increased through practices such as fertilizing and irrigating vegetation and peri-

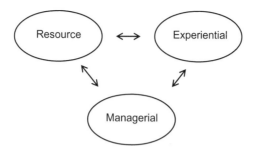

Figure 2.2. Three dimensions of carrying capacity of parks and related areas
(from Manning and Lime 1996).

odic rest and rotation of impact sites. Similarly, the quality of the recreation experience might be maintained or even enhanced in the face of increasing use by means of more even distribution of visitors, appropriate rules and regulations, provision of additional visitor facilities, and educational programs designed to encourage desirable user behavior. Thus, carrying capacity, as applied to parks and related areas, has been expanded to a three-dimensional concept by the addition of management considerations, illustrated graphically in figure 2.2.

Limits of Acceptable Change

Carrying capacity has attracted intensive focus as a research and management concept in parks and outdoor recreation. Several bibliographies, books, and review papers have been published on carrying capacity and related issues (Stankey and Lime 1973; Graefe et al. 1984; Shelby and Heberlein 1986; Stankey and Manning 1986; Kuss et al. 1990; Manning 1999), and these publications contain hundreds of citations. Despite this impressive literature base, efforts to apply carrying capacity to parks and recreation areas have often met with limited success. The principal difficulty lies in determining how much impact or change should be allowed within each of the three components that make up the carrying capacity concept: environmental resources, the quality of the recreation experience, and the extent and type of management actions.

The growing research base in parks and outdoor recreation indicates that increasing recreation use often causes impact or change. This is especially clear with regard to park resources. An early study in the Boundary Waters Canoe Area, Minnesota, for example, found that an average of 80% of groundcover vegetation was destroyed at campsites in a single season, even under relatively light levels of use (Frissell and Duncan 1965). The ecological impacts of outdoor recreation can be extensive and wide ranging, including soil compaction and erosion, trampling of vegetation, water pollution, and disturbance of wildlife, which have been summarized and synthesized in a number of studies and reports (Hammitt and Cole 1998; Cole 1987; Kuss et al. 1990; Leung and Marion 2000). Similarly, social science research has documented impacts of increasing visitor use on the quality of the recreation experience through crowding, conflict, and the aesthetic implications of resource degradation (Manning 1999; Manning and Lime 2000). Finally, research suggests that increasing recreation use can also change the management environment through development and implementation of more intensive management practices (Manning 1999; Manning and Lime 2000). Despite increasing knowledge about park and outdoor recreation use and resulting impacts, the critical question remains: how much impact or change should be allowed?

This issue is often referred to as the "limits of acceptable change" (Frissell and Stankey 1972; Stankey et al. 1985). With increasing use of parks and related areas, some change in the recreation environment—park resources, the visitor experience, the management context—is inevitable. But sooner or later the type or amount of change may become unacceptable. What determines the limits of acceptable change?

Figure 2.3 graphically illustrates a hypothetical relationship between visitor use and impacts to the resource, experience, and management components of parks. This relationship suggests that increasing recreation use can and often does cause increasing impacts in the form of damage to fragile soils and vegetation, crowding and conflicting uses, and more direct and intensive recreation management actions. However, it is not clear from this relationship at what

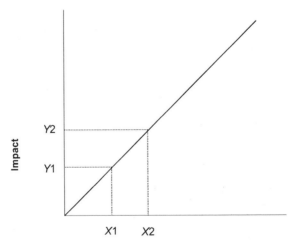

Visitor Use

Figure 2.3. Hypothetical relationship between visitor use and impact to parks and related areas (from Manning and Lime 1996).

point carrying capacity has been reached. For this relationship, *X1* and *X2* represent alternative levels of visitor use that result in corresponding levels of impact as defined by points *Y1* and *Y2*, respectively. But which of these points—*Y1* or *Y2*, or some other point along the vertical axis—represent the maximum amount of acceptable impact?

To emphasize and further clarify this issue, some studies have suggested distinguishing between descriptive and evaluative (or prescriptive) components of carrying capacity (Shelby and Heberlein 1984; Shelby and Heberlein 1986). The descriptive component of carrying capacity focuses on factual, objective data, such as the relationships shown in figure 2.3. For example, what is the relationship between the amount of visitor use and perceived crowding? The evaluative or prescriptive component of carrying capacity determination concerns the seemingly more subjective issue of how much impact or change in parks and related areas is acceptable. For example, what level of perceived crowding should be allowed?

Management Objectives and Indicators and Standards

Recent experience with carrying capacity suggests that answers to evaluative/prescriptive questions can be found through formulation of management objectives (sometimes called desired conditions) and associated indicators and standards (Lime and Stankey 1971; Frissell and Stankey 1972; Lucas and Stankey 1974; Bury 1976; P. Brown 1977; Hendee et al. 1977; Lime 1977a; Lime 1979; Stankey 1980b; Boteler 1984; Stankey et al. 1985; Stankey and Manning 1986; Graefe et al. 1990; Shelby et al. 1992; Shindler 1992; Lime 1995; Manning et al. 1995a; Manning et al. 1995b; Manning and Lime 1996; Manning et al. 1996b; Manning 1997; National Park Service 1997; Manning 2005). This approach to carrying capacity focuses on defining the level of resource protection to be maintained and the type of visitor experience to be provided. Management objectives/desired conditions are broad, narrative statements defining the type and quality of park conditions to be maintained. Indicators are more specific, measurable variables reflecting the essence or meaning of management objectives; they are quantifiable proxies or measures of management objectives. Indicators may include elements of the resource, experiential, and management environments that are important in determining the type and quality of park conditions. Standards define the minimum acceptable condition of indicator variables.

An example may help illuminate these ideas and terms. Review of the Wilderness Act of 1964 suggests that areas designated by Congress as part of the National Wilderness Preservation System are to be managed to provide opportunities for visitor "solitude." Therefore, providing opportunities for solitude is an appropriate management objective or desired condition for most wilderness areas. However, solitude is a somewhat abstract concept that is difficult to measure directly. Research on wilderness use suggests that the number of other visitors encountered along trails and at campsites is important to visitors in defining solitude. Thus, trail and camp encounters are potentially good indicator variables because they are measurable, manageable, and serve as a proxy for the management objective/desired condition of wilderness solitude. Research also suggests that wilderness visitors may have standards about how many trail and camp encounters can be experienced before opportunities for solitude decline to an unacceptable degree. For example, a number of studies suggest that wilderness visitors generally find no more than five other groups per day encountered along trails to be acceptable and wish to camp out of sight and sound of other groups. Therefore, a maximum of five encounters per day with other groups along trails and no other groups camped within sight or sound may be good standards for managing the carrying capacity of wilderness areas.

Management objectives/desired conditions and associated indicators and standards should be formulated on the basis of several considerations. In keeping with the three-dimensional model of carrying capacity illustrated in figure 2.2, these considerations can be organized into three broad categories:

- Resource: The ecological characteristics of the natural resource base of parks and related areas help determine the degree of change in the environment that results from recreation use. Some resource bases are inherently more fragile than others. These resource characteristics should be studied and may become important guides in formulating management objectives/desired conditions and associated indicators and standards.
- Experiential: The needs and wants of society are important in determining appropriate park and outdoor recreation opportunities. Studies of visitors to parks and outdoor recreation areas may suggest appropriate types and levels of outdoor recreation use and associated resource, social, and managerial impacts. Such studies should be incorporated into carrying capacity analysis and management.
- Managerial: Legal directives, agency mission statements, and other policy-related guidelines may suggest appropriate management objectives/desired conditions and related indicators and standards. Moreover, financial, personnel, and other management resources may also suggest the types and levels of park and recreation use that are feasible.

The types of information described above can be important in formulating thoughtful management objectives/desired conditions and associated indicators and standards. However, there is also a value-based element of park and recreation carrying capacity that must be addressed. Research can help illuminate the relationships between increasing use levels and change in the recreation environment—the descriptive component of carrying capacity—as illustrated in figure 2.3. Moreover, research on the standards of park visitors and other stakeholders (described more fully in chapter 5) can help inform the evaluative or prescriptive component of carrying capacity (Manning and Lawson 2002). However, some element of management judgment will ultimately be needed to integrate the resource, experiential, and managerial components of carrying capacity into informed management objectives/desired conditions and associated indicators and standards. Several carrying capacity frameworks have been developed to help guide this process.

Carrying Capacity Frameworks

The concepts and terminology described earlier have given rise to an operational definition of carrying capacity and to several frameworks for analyzing and applying carrying capacity to parks and related areas. Carrying capacity can be defined as the level and type of recreation use that can be accommodated in a park or related area without violating standards for relevant indicator variables. Carrying capacity frameworks include Limits of Acceptable Change (LAC) (Stankey et al. 1985); Visitor Impact Management (VIM) (Graefe et al. 1990); Outdoor Recreation Management Framework (Manning 1999); Visitor Experience and Resource Protection (VERP) (National Park Service 1997; Manning 2001); Carrying Capacity Assessment Process (C-CAP) (Shelby and Heberlein 1986); and Visitor Activity Management Process (VAMP) (Environment Canada and Park Service 1991). All of these frameworks incorporate the ideas about carrying capacity described above and provide a rational, structured process for conducting carrying capacity analysis and management.

The basic steps or elements of the two most widely applied carrying capacity frameworks—LAC and VERP—are shown in table 2.1. While terminology, sequencing, and other aspects may vary among these frameworks, they share a common underlying logic (Manning 2004). Core elements of these frameworks include

1. Definition of park conditions to be maintained. These conditions should be defined in terms of management objectives/desired conditions and associated indicators and standards, and should address the resource, experiential, and managerial components of parks and outdoor recreation.
2. Monitoring of indicator variables to determine if existing park conditions meet the standards that have been specified.
3. Application of management practices to ensure that standards are maintained.

As applied to parks and outdoor recreation, carrying capacity is more complex than initially envisioned (Burch 1981; Stankey 1989). Recreation-related carrying capacity includes resource, experiential, and managerial considerations, descriptive and evaluative/prescriptive components, management objectives/desired conditions and associated indicators and standards, and management judgment.

TABLE 2.1. Park and outdoor recreation carrying capacity frameworks

Limits of Acceptable Change (LAC)	Visitor Experience and Resource Protection (VERP)
Step 1. Identify area concerns and issues	Element 1. Assemble an interdisciplinary project team
Step 2. Define and describe opportunity classes	Element 2. Develop a public involvement strategy
Step 3. Select indicators of resource and social conditions	Element 3. Develop statements of primary park purpose, significance, and primary interpretive themes
Step 4. Inventory resource and social conditions	Element 4. Analyze park resources and existing visitor use
Step 5. Specify standards for resource and social indicators	Element 5. Describe a potential range of visitor experiences and resource conditions
Step 6. Identify alternative opportunity class allocations	Element 6. Allocate potential zones to specific locations
Step 7. Identify management actions for each alternative	Element 7. Select indicators and specify standards for each zone; develop a monitoring plan
Step 8. Evaluation and selection of an alternative	Element 8. Monitor resource and social indicators
Step 9. Implement actions and monitor conditions	Element 9. Take management action

It seems clear that there can be no one carrying capacity for a park or related area. Rather, carrying capacity is dependent upon how the various components of the concept are fashioned together. This complexity and apparent lack of definitiveness have caused some disillusionment. Characterizations such as "slippery" (Alldredge 1973), "elusive" (Graefe et al. 1984), and "illusive" (Becker et al. 1984) have been applied to park and recreation carrying capacity. This difficulty with carrying capacity seems to be borne out in surveys of park and wilderness managers (Washburne 1981; Washburne and Cole 1983; Manning et al. 1996c; Abbe and Manning 2005). Even though many managers suspect that recreational use of their areas has exceeded carrying capacity, they have not yet established such carrying capacities.

The weaknesses and shortcomings of carrying capacity have been noted by a number of writers (Buckley 1999; Lindberg and McCool 1998; Lindberg et al. 1997). Several point out that the term may imply a single "magic number" for each park and recreation area, and that this, of course, is misleading and obscures the role of management judgment (Bury 1976; Washburne 1982). For this reason, a stronger emphasis on management objectives has been suggested by some as an alternative to carrying capacity (Becker and Jubenville 1982; Jubenville and

Becker 1983; Stankey et al. 1984). Similarly, it has been noted that analyses of carrying capacity can ignore the ability of management to affect the amount of use that can be accommodated; the term *design capacity* has been suggested as an alternative to carrying capacity (Godin and Leonard 1977b).

Others have argued that the very term *carrying capacity* seems to imply an undue emphasis on use limitations (Washburne 1982; Burch 1984; Stankey et al. 1984). These writers argue that a number of management practices might be used to meet management objectives aside from use limitations, which may often be the least preferred alternative. Moreover, while management objectives for some areas may well set relatively low carrying capacities and thus ultimately require use limits, other areas will properly have relatively high carrying capacities and may not require use limits. In a similar vein, it has been noted that recreation-caused change is not inherently undesirable (Stankey 1974). In fact, use of the more neutral word "change" has been suggested, as opposed to "impacts," "damage," or other value-laden terms, since judgment about the relative desirability of change can only be made in relationship to management objectives.

Finally, even the author of the original conceptual analysis of recreation carrying capacity has suggested that borrowing the term from range and wildlife management may not have been a wise choice (J. A. Wagar 1974). The close association between carrying capacity and resource or ecological considerations in the historical sense tends to divert attention from the equally important experiential and managerial concerns that must be a part of carrying capacity as applied to parks and outdoor recreation.

All of these points are valid criticisms. However, the term carrying capacity is deeply entrenched in the field of parks and outdoor recreation (and in environmental management more broadly), and recent legislation and institutional directives have even made carrying capacity a formal part of park and outdoor recreation management (Manning et al. 1996d). For example, recent revisions to NPS policies require that management plans for all units of the national park system address carrying capacity. More important, carrying capacity represents a vital issue of growing urgency in parks and outdoor recreation. It is a specific manifestation of the tragedy of the commons (as described in chapter 1) and requires informed and explicit management action.

Despite its shortcomings, the term carrying capacity is likely to remain an important part of the field of parks and protected areas. Moreover, carrying capacity can be useful as a park and outdoor recreation management concept when viewed in proper perspective—as an organizing framework for analyzing, defining, and managing appropriate park and outdoor recreation conditions. The carrying capacity frameworks developed in the literature and their successful application in the field (as illustrated in the part 3 of this book) suggest that carrying capacity—in its contemporary interpretation—can be a useful concept for managing parks and related areas.

CHAPTER 3

Indicators and Standards

We want to maximize good per person, but what is good?

Chapter 2 described the way in which indicators and standards have emerged as a central focus of contemporary carrying capacity frameworks for parks and protected areas. *Indicators* are measurable, manageable variables that help define the quality of parks and outdoor recreation areas. *Standards* define the minimum acceptable condition of indicator variables. Carrying capacity can be managed by means of monitoring indicator variables and implementing management actions to ensure that standards are maintained. This chapter discusses indicators and standards in more detail. Desirable characteristics of both indicators and standards are described, and examples of indicators and standards for parks and related areas are compiled and discussed.

Characteristics of Good Indicators

Several studies have explored characteristics that define good indicators (Schomaker 1984; Stankey et al. 1985; Merigliano 1990; Whittaker and Shelby 1992; National Park Service 1997; Vaske et al. 2000). These characteristics can be used to further understand the role of indicators and standards in park and outdoor recreation management and to assist in evaluation and selection of potential indicator variables. Characteristics of good indicators include the following:

- Specific: Indicators should define specific rather than general conditions. For example, "solitude" would not be a good indicator because it is too general.

"The number of other groups encountered per day along trails" would be a better indicator variable.

- Objective: Indicators should be objective rather than subjective. That is, indicator variables should be measured in absolute, unequivocal terms. Variables that are subjective, expressed in relative terms, or subject to interpretation make poor indicators. For example, "the number of people at one time at Delicate Arch" is an objective indicator because it refers to an absolute number that can be readily counted and reported. However, "the percentage of visitors who feel crowded at Delicate Arch" is a subjective indicator because it is subject to interpretation by visitors—it depends on the types of visitors making the judgment, the behavior of other visitors, and other mediating influences.

- Reliable and repeatable: An indicator is reliable and repeatable when measurement yields similar results under similar conditions. This criterion is important because monitoring of indicator variables is often conducted by more than one person; monitoring should take place at regular intervals over a long period of time.

- Related to visitor use: Indicators should be related to at least one of the following attributes of visitor use: level of use, type of use, location of use, or behavior of visitors. A major role of indicators is to help determine when management action is needed to control the impacts of visitor use. Thus, there should be a correlation between visitor use and indicator variables.

- Sensitive: Indicators should be sensitive to visitor use over a relatively short period of time. As the level of use changes, an indicator should respond in roughly the same proportional degree. If an indicator changes only after impacts are substantial, it will not serve as an early warning mechanism, allowing managers to react in a timely manner.

- Manageable: Indicators should be responsive to, and help determine the effectiveness of, management actions. The underlying rationale of indicators is that they should be maintained within prescribed standards. This implies that they must be manageable.

- Efficient and effective to measure: Indicators should be relatively easy and cost-effective to measure. Indicators must be monitored on a regular basis. Therefore, the more expertise, time, equipment, and staff needed to take such measurements, the less desirable a potential indicator may be.

- Integrative or synthetic: There are potentially many management objectives/ desired conditions to be achieved and maintained in parks and related areas. As noted in chapter 2, these might apply to park resources, the quality of the visitor experience, and the type and level of management. However, it is probably impractical to monitor large numbers of indicator variables. For this reason, integrative or synthetic indicators—variables that are proxies for more than one component of parks and protected areas—are especially useful. For example, an indicator of trail or campsite impacts may be useful as a measure of resource conditions as well as an indicator of the aesthetic dimension of the quality of the visitor experience. Or an indicator of the level of visitor use may be useful as a measure of crowding and associated resource and social impacts.

- Significant: Perhaps the most important characteristic of indicators is that they help define the quality of park resources and the visitor experience. This is inherent in the very term indicator. It does little good to monitor the condition of a variable that is unimportant in defining the quality of park resources and associated experiences.

It may be useful to incorporate these characteristics within a matrix, as shown in figure 3.1, for the purpose of evaluating potential indicators. Potential indicator variables can be arrayed along the horizontal axis of the matrix and the characteristics of good indicators arrayed along the vertical axis. Potential indicators can then be rated as to how well they meet those characteristics. Indicators that receive the highest aggregate ratings may have the greatest value in measuring and managing carrying capacity.

Potential indicators	Criteria for good indicators								
	Specific	Objective	Reliable and repeatable	Related to visitor use	Sensitive	Manageable	Efficient and effective to measure	Integrative or synthetic	Significant
Indicator 1									
Indicator 2									
Indicator 3									
Indicator 4									
Indicator 5									
Indicator . . .									

Figure 3.1. Evaluation matrix for selecting indicators.

Potential Indicator Variables

A growing body of research has focused on identifying potential indicators for a variety of park and recreation areas and activities. This research has been aimed at determining variables important to visitors in defining the quality of park resources and the recreation experience. Potential indicators identified in these studies are compiled in appendix A.

These studies have addressed a variety of park and recreation areas and activities and utilized several study methods, including open- and closed-ended questions and surveys of visitors, interest groups, managers, and scientists (these research methods are described more fully in chapter 4). However, several general conclusions might be derived from these findings.

First, it is apparent that potential indicators can be wide ranging. It may be useful to employ the three-fold framework of carrying capacity (resource, experiential, and managerial) described in chapter 2 when thinking about potential indicators. All of the indicator variables in appendix A can be classified into these resource, experiential, and managerial components.

Second, most of the studies on indicators have found some variables more important than others. For example, litter and other signs of use impacts appear to be universally important. Management-related impacts (e.g., signs, presence of rangers) appear to be less important. Level of visitor use appears universally important, but how this is manifested may be even more significant. For example, the type of visitor encountered (e.g., hikers encountering bikers or stock users, floaters encountering motor boaters) may be just as important as the number of encounters (Manning et al. 2000). In the context of backcountry or wilderness, the number of other groups encountered may be important, while in frontcountry the number of people seen at one time at attraction sites may be more operative. In other park contexts, the impacts of level of use may be manifested in terms of waiting times, competition for access, or in other ways.

Third, visitors to wilderness and related sites may be generally more sensitive to a variety of potential indicators than visitors to more highly used and developed areas or sites. However, research may have simply not yet identified and studied indicators that are most important to visitors in more highly used areas.

Characteristics of Good Standards

Several studies have explored characteristics that might define good standards (Schomaker 1984; Brunson et al. 1992; Whittaker and Shelby 1992; National Park Service 1997; Vaske et al. 2000). To the extent possible, standards should incorporate the following characteristics:

- Quantitative: Standards should be expressed in a quantitative manner. Since indicators are specific and measurable variables, standards can and should be expressed in an unequivocal way. For example, if an indicator is "the number of encounters with other groups per day on the river," then the standard might be "an average of no more than three encounters with other groups per day on the river." In contrast, "low numbers of encounters with other groups per day on the river" would be a poor standard because it does not specify the minimum acceptable condition in unambiguous terms.

- Time- or space-bounded: Incorporating a time- or space-bounded element into a standard expresses both how much of an impact is acceptable and how often or where such impacts can occur. It is often desirable for standards to have a time period associated with them. This is especially relevant for crowding-related issues. For instance, in the above example, the standard for encounters with other groups on the river was expressed in terms of "per day." Other time-bounded qualifiers might include "per night," "per trip," "per hour," or "at one time," depending upon the circumstances.

- Expressed as a probability: In many cases, it will be advantageous to include in the standard a tolerance for some percentage of the time that the desired condition may not be met. For example, a standard might specify that hikers will have "no more than three encounters with other groups per day along trails for 90% of the days in the summer-use season." The 90% probability of conditions meeting or exceeding the standard allows for 10% of the time that random or unusual events might prevent management from reasonably maintaining these conditions. This allows for the complexity and randomness inherent in park-use patterns. In the example of encounters along a trail, several hiking parties might depart from a trailhead at closely spaced intervals on a given day. These groups are likely to encounter each other on the trail several times during the day. On another day, the same number of groups might depart from the trailhead at widely spaced intervals and thereby rarely encounter each other. Similarly, it might be wise to incorporate a tolerance in standards for peak-use days, holiday weekends, or other days of exceptionally high visitation. The amount of tolerance needed depends on the unpredictability of each individual situation and the degree to which management can consistently control conditions.

- Impact-oriented: Standards should focus directly on the impacts that affect the quality of park resources and the visitor experience, not the management action used to keep impacts from violating the standards. For example, an appropriate standard might be "no more than ten encounters with other groups on the river per day." This could be a good standard because it focuses directly on the impact that affects the quality of the visitor experience—the number of other groups encountered. Alternatively, "a maximum of twenty groups per day floating the river" would not be as good a standard because it does not focus as directly on the impact of concern; visitors experience encounters with other groups more directly than they experience total use levels. Basing standards on management actions rather than on impacts can also limit consideration of the potential range of useful management practices (management practices to maintain standards are described in part 4 of this book). For example, limiting the number of boats to twenty per day might be used to ensure ten or fewer encounters per day, but other actions, such as more tightly scheduling launch times, could also ensure an acceptable encounter rate and could be less restrictive on the level of visitation to the river.

- Realistic: Standards should generally reflect conditions that are realistically attainable. Standards that limit impacts to extremely low levels may set up unrealistic expectations in the minds of visitors and the general public, may be politically infeasible, and may unfairly restrict visitor use to very low levels.

Potential Standards

An increasing number of studies have been conducted to help define standards for parks and related areas. Most of these studies have incorporated normative theory and methods as described in chapter 5. These studies have addressed a variety of park and outdoor recreation areas and potential indicators. They have also used alternative question formats and wording, different response scales, and other methodological variations. Potential standards reported in these studies are compiled in appendix B. Several general conclusions might be derived from this growing body of literature.

First, standards can be measured for a variety of potential indicators. While many studies have addressed encounter and other crowding-related indicator variables, other studies have measured standards for widely ranging variables. Standards have been measured for a variety of resource, experiential, and managerial indicators, representing all three of the components of carrying capacity described in chapter 2.

Second, visitors tend to report standards more often in wilderness or backcountry situations than in frontcountry or more developed areas. Moreover, there tends to be more agreement or consensus about wilderness/backcountry-related standards. (This issue is often called *crystallization* and is discussed more fully in chapter 5). For example, standard deviations of encounter standards for floaters on three western rivers were found to increase as the recreation opportunity described moved from "wilderness" to "semi-wilderness" to "undeveloped recreation" (Shelby 1981). Moreover, the percentage of floaters on the New River, West Virginia, who reported a series of encounter-related standards decreased across a similar spectrum of recreation opportunities (Roggenbuck et al. 1991).

Third, standards tend to be lower (or less tolerant) in wilderness or back-country areas than in frontcountry or more developed areas. For example, visitors to wilderness areas tend to want to camp out of sight and sound of other groups while visitors to developed campgrounds are tolerant of relatively large numbers of other groups.

Fourth, there may be some consistency in standards within similar types of parks and outdoor recreation areas or opportunities. For instance, a study of visitor standards for a variety of potential indicators found broad agreement across the four geographically diverse wilderness areas included in the study (Roggen-buck et al. 1993). Moreover, a number of studies suggest that standards for encountering other groups along trails during a wilderness experience are quite low (about five or fewer per day) and that many wilderness visitors prefer to camp out of sight and sound of other groups. More studies of park and outdoor recreation-related standards will allow more definitive tests of the extent to which such standards might be generalized across areas.

Fifth, standards of visitors can vary from those of managers. For example, a study of standards for wilderness campsite impacts found that visitors reported more restrictive standards regarding the presence of fire rings and tree damage than did managers (Martin et al. 1989). However, managers reported more restrictive standards for bare ground impacts.

PART II

Research to Support Application of Carrying Capacity

It seems clear from part 1 that carrying capacity of parks and protected areas is a serious issue of increasing urgency. As a specific manifestation of the tragedy of the commons, carrying capacity must be addressed through social action—"mutual concern, mutually agreed upon." In accordance with evolving interpretation of carrying capacity as a normative concept, and associated frameworks developed to analyze and manage carrying capacity, management objectives (or desired conditions) must be developed for parks and related areas and ultimately expressed in the form of indicators and standards. Indicators must be monitored and management actions taken to ensure that standards are maintained.

While the process outlined in these carrying capacity frameworks is clear, the elements of this process can be challenging and sometimes even contentious. What are desired conditions of parks and related areas? What are good indicators of these desired conditions? What are minimum acceptable standards for indicator variables? How can indicators be monitored in efficient, effective ways? What management actions are acceptable and effective at maintaining standards?

Answers to these and related questions will always require some element of management judgment. However, such judgments should be as informed as possible (Manning and Lawson 2002). Recent research has identified and tested a number of scientific approaches that can inform carrying capacity analysis and management. Just as carrying capacity itself is an interdisciplinary concept (involving resource, experiential, and managerial components), these research approaches have drawn on theory and methods from a number of disciplines, including sociology, psychology, economics, ecology, statistics, landscape architecture, business management, and computer science. The five chapters that comprise part 2 describe and illustrate these research approaches, including qualitative and quantitative surveys, normative theory and methods, visual research, tradeoff analysis, and computer-based simulation modeling.

CHAPTER 4

Identifying Indicators for Parks and Protected Areas

What shall we maximize?

As discussed in chapter 3, indicators serve as quantifiable measures of management objectives/desired conditions for parks and protected areas. They can (and in most cases should) apply to both park resources and the quality of the visitor experience, including the type and intensity of management practices. Indicators are measurable, manageable variables that help define the quality of parks and related areas. Several approaches can be used to help identify potential indicator variables.

Qualitative Interviews

Qualitative interviews are open-ended, in-depth discussions with respondents; they are generally characterized by a series of questions that encourage respondents to think about and discuss their opinions or experiences (Tashakkori and Teddlie 1998; Patton 2002). Interviews are guided by a structured series of questions, but interviewers are permitted to ask other clarifying or exploratory questions. This survey method is termed qualitative because study findings are designed to describe the range of opinions or experiences in a population as opposed to estimating their quantitative distribution throughout that population. Qualitative surveys are often conducted using purposive rather than representative sampling to help ensure that as full a range of responses as possible is derived. Purposive sampling might be designed on the basis of type of respondent or diversity of activities or sites within a park. Interviews are usually recorded, transcribed and coded to identify important themes and, ultimately, indicators (Patton 2002; Miles and Huberman 1994; Coffey and Atkinson 1996).

As an example of the qualitative approach, interviews were conducted with visitors and other stakeholders in conjunction with application of the Visitor Experience and Resource Protection (VERP) framework (as described in chapter 2) to Arches National Park, Utah. The purpose of the interviews was to help identify indicators of the quality of the visitor experience, including the ways in which the condition of park resources affected the visitor experience. A semi-structured interview script was developed that asked a series of probing, open-ended questions about what respondents felt were the most important qualities or characteristics of the visitor experience at Arches. Interviews were conducted in the park with 112 visitors at seven sites. In addition, ten focus group sessions were also conducted with a total of eighty-three participants. Participants included park staff, visitors who participated in the park's interpretive programs, and residents of the local community.

Responses were initially coded into ninety-one categories and then grouped into several major themes or subject matter classes. Themes that best met the characteristics of good indicators as described in chapter 3 (e.g., measurable, manageable, related to visitor use, important in affecting the quality of the visitor experience) included crowding at attraction sites and along trails, visitors walking off trails and damaging soils and vegetation, and vehicle traffic on park roads.

Quantitative Surveys

Quantitative surveys are generally characterized by a series of close-ended questions with defined response scales. This survey method is termed quantitative because it is designed to measure the distribution of responses (e.g., characteristics of respondents, opinions, experiences) throughout a population. Quantitative surveys are conducted using representative sampling methods incorporating an element of randomness. Study findings are coded, analyzed, and reported using mathematical and statistical procedures.

As an example, the program of research at Arches National Park (described earlier) conducted a second visitor survey that incorporated a battery of close-ended questions addressing fourteen potential indicator variables. These potential indicators were developed from the initial qualitative survey, literature review, and the judgment of park staff and planning team members. The survey was administered to representative samples of visitors at seven sites within the park. Respondents were asked to rate the importance of each potential indicator on a five-point scale that ranged from one ("very unimportant") to five ("very important"). For park visitors as a whole, the most important indicators were vandalism, litter, inappropriate behavior of visitors (including walking off trails), damage to soils and vegetation, visitor-caused noise, and crowding at attraction sites and along trails. However, there were some differences among study sites. For example, visitors to developed areas in the park rated the indicator of visitors walking off trails as more important than did visitors to the backcountry.

Qualitative and quantitative survey approaches can be employed in a complementary way. As with the example at Arches National Park, qualitative interviews can be used to help identify the range of potential indicators, and follow-up quantitative surveys can help determine the relative importance of those indicators across representative samples of park visitors or other relevant publics.

Importance Performance Analysis

An "importance-performance" framework might also be used as an aid to formulating indicators (Mengak et al. 1986; Hollenhorst and Stull-Gardner 1992; Hollenhorst et al. 1992; Hollenhorst and Gardner 1994). Using quantitative surveys in the context of this framework, respondents are first asked to rate the importance of potential indicator variables, and these results are plotted along a vertical axis as shown in figure 4.1. Second, visitors are asked a series of norma-

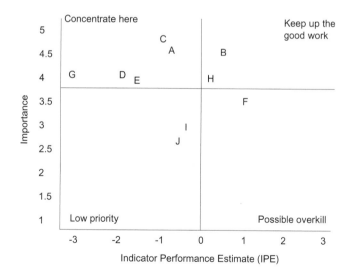

A = Number of parties of people I see each day. B = Number of large parties (more than six people) I see each day. C = Number of parties camped within sight and sound of my campsite. D = Number of parties that walk past my campsite each night. E = Number of visible places I see each day where people have camped. F = Number of horse parties encountered each day. G = Percentage of vegetation loss and bare ground seen each day. H = Number of fire rings. I = Number of signs seen each day. J = Number of culverts seen each day.

Figure 4.1. Importance-performance analysis (from Hollenhorst and Gardner 1994).

tive questions (described in chapter 5) regarding standards for each indicator variable. These data are then related to existing park conditions and plotted on a horizontal axis. The resulting data provide a graphic representation of the relationship between importance and performance of indicator variables, and where management action should be directed. The data in figure 4.1, for example, are derived from a survey of visitors to the Cranberry Wilderness, West Virginia, and suggest that indicator variable A ("number of parties of people I see each day") is important to visitors, but that visitors currently see more parties per day than their normative standard (Hollenhorst and Gardner 1994). These findings suggest that managers should concentrate their attention on this indicator variable.

Threats Matrix

A park and protected area "threats matrix" is another framework that might be applied to help identify potential indicators (Leopold et al. 1971; Manning and Moncrief 1979; Cole 1994). A matrix model of park and outdoor recreation impacts can be created by arraying important attributes of outdoor recreation to form the rows of a matrix, and arraying potential threats to those attributes as the columns of the matrix. Each cell within the matrix represents the various impacts that each threat causes to each attribute. An example of such a matrix is shown in figure 4.2. This example was developed to determine the significance

Wilderness Threats

Attributes of wilderness character	Recreation	Livestock	Mining	Fire	Exotic species	Water projects	Atmospheric pollutants	Adjacent lands
Air	1	1	1	2	1	1	4	3
Aquatic systems	4	3	3	4	4	3	4	3
Rock/ landforms	1	2	2	1	1	2	1	1
Soils	3	3	2	5	2	2	4	2
Vegetation	3	3	2	5	4	3	4	2
Animals	4	2	2	4	3	2	2	4
Ecosystems/ landscapes	2	3	2	5	3	2	4	5
Cultural resources	3	2	2	2	1	1	1	1
Wilderness experiences	4	3	2	3	2	2	2	3

Figure 4.2. Wilderness threats matrix. Matrix values are significance ratings for the impacts of each potential threat on each wilderness attribute for all wilderness areas in the U.S. Forest Service's Northern Region. Ratings range from 1 (low) to 5 (high). (From Cole 1994.)

of threats to wilderness areas within one region of the national forest system (Cole 1994). This example applies to wilderness very broadly but can be developed more specifically for parks and outdoor recreation. Such a matrix can be useful as a means of identifying potential indicators (important attributes of parks and related areas that are impacted by potential threats), and the extent to which such indicator variables are threatened and, therefore, need monitoring and management attention.

Other Approaches

Other research and management approaches might be used to develop insights into potential indicators. Normative theory and methods used to measure standards for indicator variables can suggest the importance or "salience" of potential indicators as a function of how strongly respondents rate the acceptability of a range of recreation-related resource and social impacts. This approach is described more fully in chapter 5.

Stated choice survey and related statistical methods can also be used to measure the relative importance of potential indicator variables. This approach is described more fully in chapter 7 and can be important because it asks respondents to reveal preferred tradeoffs among potentially competing or conflicting indicators.

Of course, ecological research and related inventorying and monitoring approaches can also be useful in identifying resource-related indicators. A range of natural science-based research methods have been adapted and applied to study recreation-related impacts in parks and protected areas, and findings from this body of work have been synthesized in recent reports (Hammitt and Cole 1998; Leung and Marion 2000; Buckley 2004).

Finally, potential indicators can be identified though the planning and management process. For example, inventory and assessment of park conditions can identify significant and/or endangered park resources that warrant representation by indicator variables. Moreover, review of pertinent legislation or agency policy may identify selected types of park resources or visitor experiences that should be provided and maintained. The Wilderness Act (1964) and its focus on solitude, as described in chapter 2, is an example. Finally, contemporary public planning typically offers opportunities for the public to participate in this process through public meetings and other methods, and this may provide insights into potential indicator variables.

CHAPTER 5

Normative Standards for Indicator Variables

The only kind of coercion I recommend is mutual coercion,
mutually agreed upon by the majority of the people affected.

Norms are a theoretical construct that have a long tradition and are widely used in the discipline of sociology and the social sciences more broadly. As the word suggests, norms represent what is considered "normal" or generally accepted within a cultural context. For example, people in most Western societies pass one another on the right when they meet on sidewalks. In a more technical sense, norms are cultural rules that guide behavior. Moreover, such behavior is a function of a sense of obligation to abide by the norm and a belief that sanctions (rewards or punishments) may be forthcoming, depending on whether or not norms are followed (Grasmick et al. 1993; Heywood 2002; Vaske and Whittaker 2004). It is this sense of obligation and associated sanctions that make norms different from, and potentially more powerful than, *attitudes*. Attitudes are positive or negative evaluations of behavior, while norms define what behavior *should* be. Sanctions associated with norms can range from informal and internally imposed (e.g., feeling good or guilty) to formal and externally imposed (e.g., public recognition or being publicly ostracized). When norms apply to behaviors that are important to society and for which there is wide agreement, they can ultimately be codified into administrative rules and regulations, public policy, or even law (e.g., vehicles must be driven on the right side of the road).

Normative theory has developed along three basic lines (Vaske and Whittaker 2004). One branch of normative theory addresses the variables that activate norms or bring them into focus (Cialdini et al. 1991; Cialdini et al. 1990). A second branch of theory deals with how completely attitudes and norms ultimately direct behavior (Ajzen and Fishbien 1980; Fishbein and Ajzen 1975). A third branch of normative theory and methods—structural characteristics models—has special application to carrying capacity, and formulation of standards in particular. This work has been based largely on development of the *return potential model* (Jackson 1965). In the context of carrying capacity, this model works by asking survey respondents (e.g., park visitors, residents of surrounding communities, the general public) to evaluate the acceptability (or other evaluative dimension, as discussed later in this chapter) of a range of recreation-related impacts to park resources or the quality of the visitor experience. Resulting data are generally graphed so that impacts are displayed on the horizontal axis and evaluations are displayed on the vertical axis. The resulting line connecting the evaluation scores is often called an *impact acceptability curve* or simply a *norm curve*. A hypothetical norm curve is shown in figure 5.1. In this case, a sample of park visitors might have been asked to rate the acceptability (using a nine-point response scale) of encountering a range of other groups per day while hiking along a park trail.

Norms can be measured for both individuals (personal norms) and groups (social norms). As the terms suggest, personal norms are measures of the standards or evaluations of individuals, while social norms represent shared standards or the evaluations of a group. Social norms are measured by aggregating the evaluation data for members of a group. The resulting line (as illustrated in figure 5.1) is often called a *social norm curve.*

Structural characteristics models of norms can be especially useful in helping to formulate carrying capacity–related indicators and standards that are vital components of carrying capacity frameworks, as described in chapters 2 and 3.

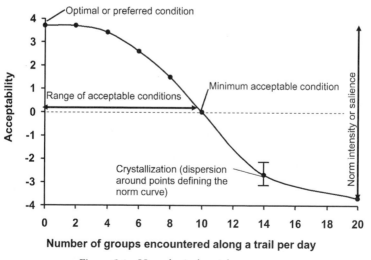

Figure 5.1. Hypothetical social norm curve.

If park visitors or other interest groups have shared norms for the condition of park resources and/or the visitor experience, then such norms can be studied and used as a basis for formulating standards. In this way, carrying capacity can be determined and managed more effectively.

This approach to normative theory and methods has been increasingly applied to the field of park and outdoor recreation management, and has addressed elements of the resource, social, and management components of carrying capacity (Shelby and Heberlein 1986; Vaske et al. 1986; Whittaker and Shelby 1988; Shelby et al. 1988a; Patterson and Hammitt 1990; Williams et al. 1991; Vaske et al. 1996; Manning et al. 1996a; Manning et al. 1996b; Manning 1997; Manning et al. 1998a; Jacobi and Manning 1999). This work has also been extended to other applied areas of environmental and natural resources management, including wildlife management (Zinn et al. 1998; Zinn et al. 2000; Wittmann et al. 1998; Whittaker 1997), fire management (Bright et al. 1993; Kneeshaw et al. 2004), and minimum stream flows (Shelby and Whittaker 1995).

Application of normative theory and methods to parks, carrying capacity, and natural resources management issues more broadly involves extension of normative theory and methods as originally conceived (Roggenbuck et al. 1991, Shelby and Vaske 1991; Vaske and Whittaker 2004). Many of these applications address resource and social conditions, not behavior. Moreover, unlike behavior, resource and social conditions do not appear to be subject to sanctions, nor do they entail an explicit notion of obligation on the part of individuals. However, visitor-caused impacts to park resources and the quality of the recreation experience are a direct *consequence* of visitor behavior. Moreover, the decision to manage such impacts in relation to socially acceptable levels represents *institutional* behavior of management agencies. These agencies have an obligation to manage parks and related areas to meet the needs of society, and these agencies are ultimately subject to *sanctions* (e.g., public disapproval, legal challenge) if they are perceived to fail to live up to this obligation.

This extension of normative theory allows application of norms to carrying capacity and to the tragedy of the commons more broadly. It will be remembered from chapter 1 that the tragedy of the commons arises in instances when it is rational for an individual to use (and impact) a resource because the marginal benefit is greater than the marginal cost to that individual. However, for society as a whole, marginal costs will ultimately exceed marginal benefits (thus, carrying capacity is exceeded and the tragedy of the commons materializes). In this circumstance, social norms can help define resource and social conditions (expressed in terms of indicators and standards) desired by society. These social norms and associated standards oblige agencies to manage parks and related areas to maintain these conditions, thus constraining the behavior of individuals (through, for example, encouraging low-impact activities or limiting use) to limit impacts to park resources and the quality of the visitor experience. In this way, standards derived from social norms constitute the "mutual coercion, mutually agreed upon" needed to avert the tragedy of the commons and its specific manifestation of carrying capacity of parks and related areas.

Measuring Norms

As outlined earlier, the hypothetical social norm curve illustrated in figure 5.1 is derived from a series of questions that might ask respondents to judge the acceptability of meeting a range of other groups along a trail in a day. The social norm curve is constructed from the mean (or median) acceptability ratings for the sample as a whole and can simply connect these points with a series of straight lines or, as represented in the illustration, can be a regression line, which serves to interpolate between points and "smooth" the curve. A "real" social norm curve is shown in figure 5.2. In this example, a representative sample of wilderness hikers in Zion National Park, Utah, were asked to rate the acceptability (on a nine-point response scale) of encountering between zero and sixteen groups of hikers per day along park trails. Average (mean) acceptability ratings were used to construct the resulting social norm curve.

Social norm curves have several potentially important features or characteristics that can contribute to their interpretation and usefulness, as illustrated in

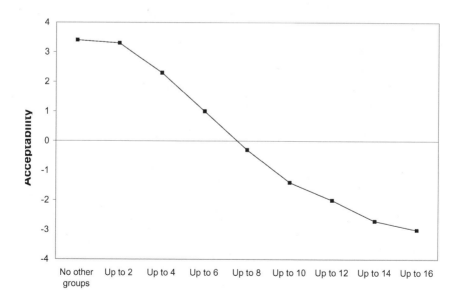

Number of other groups encountered per day

Figure 5.2. Social norm curve for groups encountered per day for wilderness hikers in Zion National Park.

figure 5.1. First, all points along the curve above the neutral point on the accept-ability scale—the point on the vertical axis where aggregate evaluation ratings fall out of the acceptable range and into the unacceptable range—define the *range of acceptable conditions.* All of the conditions represented in this range are judged to meet some aggregate level of acceptability. The *optimum* or *preferred* condition is defined by the highest point on the norm curve. This is the condition that received the highest rating of acceptability from the sample as a whole. The *minimum acceptable condition* is defined as the point at which the norm curve crosses the neutral point of the acceptability scale. This is the point at which aggregate ratings of the condition of the indicator variable fall out of the acceptable range and into the unacceptable range. *Norm intensity* or *norm salience*—the strength of respondents' feelings about the importance of a potential indicator—is sug-gested by the amplitude of the curve or the distance of the norm curve above and below the neutral point of the evaluation scale. The greater this distance, the more strongly respondents feel about the indicator or the condition being meas-ured. High measures of norm intensity or salience suggest that a variable may be a good indicator because respondents feel it is important in defining the quality of park resources or the recreation experience. *Crystallization* of the norm con-cerns the amount of agreement or consensus about the norm. It is usually meas-ured by standard deviations or other measures of variance of the points that describe the norm curve. The less variance or dispersion of data about those points, the more consensus there is about social norms.

Norms can also be measured using a shorter, open-ended question format by asking respondents to report the maximum level of impact that is acceptable or preferable. (A number of evaluative dimensions might be used in normative questions, and this issue is discussed later in this chapter.) For example, a repre-sentative sample of backpackers in Yosemite National Park, California, were asked to report the maximum number of groups per day they preferred to encounter along park trails. The resulting frequency distribution of responses is shown in table 5.1 and figure 5.3. A plurality of respondents reported that they

TABLE 5.1. Frequency distribution of preferred
number of encounters/day on the trails in the wilderness
portion of Yosemite National Park, CA

Number of trail encounters	Frequency	Cumulative %
5 or less	265	36.9
6–10	192	63.6
11–15	76	74.2
16–20	73	84.4
21–25	37	89.6
26–30	25	93.1
>30	8	100.0

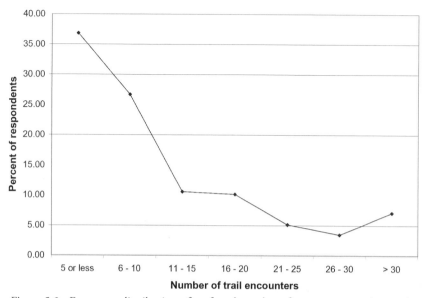

Figure 5.3. Frequency distribution of preferred number of encounters per day on the trails in the wilderness portion of Yosemite National Park.

preferred to encounter no more than five groups per day, and the vast majority of respondents agreed that it was preferable to encounter no more than ten groups per day. This open-ended question format is designed to be less burdensome to respondents, but it also yields less information. Alternative question formats for measuring norms are addressed more fully later in this chapter.

Theoretical and Methodological Issues

A considerable body of literature has emerged on applications of normative theory and methods to standards for parks and related areas. This work has addressed a variety of potential indicator variables that represent all three components of the concept of carrying capacity: resource, experiential, and managerial. Examples include (1) the maximum amount of soil erosion on trails, (2) the maximum number of groups encountered along trails, and (3) the appropriate level of trail development, respectively.

Application of normative research has raised a number of theoretical and methodological issues that can help guide this research, its interpretation, and its ultimate incorporation into park planning and management. For example, research has found that most respondents are able to report or specify a normative standard for most indicator variables included in most studies (Donnelly et al. 2000). This issue is sometimes referred to as *norm prevalence* (Kim and Shelby

1998). For example, 87% of canoeists in the Boundary Waters Canoe Area Wilderness, Minnesota, reported a norm for the maximum acceptable number of other parties seen per day on the lake or river where they spent the most time (Lewis et al. 1996a; Lewis et al. 1996b). A comparative analysis of data from over fifty study contexts found that norm prevalence (the percentage of respondents who reported a normative standard) averaged 70%. Norm prevalence was highest in the context of wilderness or backcountry settings (as opposed to frontcountry) and for encounters among visitors that included some element of conflict (e.g., canoeists and motor boaters).

However, a few studies have found relatively low norm prevalence. For example, a study of floaters on the New River, West Virginia, found that only 29% to 66% of respondents reported a norm for several indicator variables under three alternative types of recreation opportunities (Roggenbuck et al. 1991). Other visitors chose one of two other response options, indicating that the potential indicator did not matter to them, or that it did matter, but they couldn't specify a maximum amount of impact acceptable. Reasons as to why visitors may not be able to report norms are discussed later in this chapter.

As suggested in chapter 3, there may be some emerging consistency in norms within similar types of parks and recreation areas or opportunities. For instance, the data on normative standards compiled in appendix B from a number of studies suggest that norms for encountering other groups while hiking during a wilderness experience are quite low (about five or fewer) and that many wilderness visitors prefer to camp out of sight and sound of other groups.

Research suggests that norms generally fall into one of three categories or types: no-tolerance, single-tolerance, and multiple-tolerance. For example, a study of boaters on the Deschutes River, Oregon, measured norms for a number of potential indicators and found all three types of norms, as shown in figure 5.4 (Whittaker and Shelby 1988). The social norm curve for human waste represents a no-tolerance norm: the majority of respondents report that it is never acceptable to see signs of human waste along the river. Other indicators for which no-tolerance norms were reported included selected types of discourteous behavior, and jet boat encounters for non–jet boaters. No-tolerance norms tend to be characterized by a mode at zero impact, high intensity, and high crystallization.

The social norm curve for time in sight of others represents a single-tolerance norm: the vast majority of respondents were willing to tolerate some time in sight of others but were unwilling to accept such impact beyond a certain level (two hours out of four in sight of others). Other indicators for which single-tolerance norms were reported included jet boat encounters for jet boaters, launch waiting times, angling disturbances, fishing competition, camp sharing, and camp competition. Single-tolerance norms tend to be characterized by a mode at some level of impact greater than zero and a sharp decline in the percentage of respondents reporting tolerances for impacts greater than the modal value.

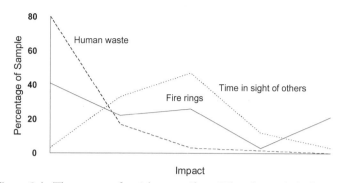

Figure 5.4. Three types of social norms (from Whittaker and Shelby, 1988).

The social norm curve for fire-ring impacts represents a multiple-tolerance norm: multiple "peaks" along the norm curve indicate there are at least two groups of respondents with distinctly different normative standards for this indicator.

Several studies have focused attention on the issue of norm salience. Early in this chapter, salience was defined as the importance of potential indicators in determining the quality of park resources or the recreation experience. The issue of salience may help explain why some respondents do not report personal norms (Shelby et al. 1996; Donnelly et al. 2000). When relatively large percentages of respondents do not report norms, it may be that the indicator or impact under study is not important in determining the quality of park resources or the recreation experience. Several studies are suggestive of the role of salience in recreation-related norms. As noted earlier, relatively low numbers of floaters on the New River, West Virginia, reported norms for encounter-related indicators when compared to other river recreation studies (Roggenbuck et al. 1991). However, the New River is a relatively high-use area and encounter-related indicators may be less important or salient in this context. This reasoning is supported by other studies, as described earlier, which have found that higher percentages of respondents reported norms for wilderness or backcountry areas than for frontcountry areas. Many of the indicators addressed in these studies are encounter-related and may simply be less important or salient in frontcountry than in backcountry.

A closely related issue concerns how indicators or impacts are manifested and ultimately perceived by park visitors. Measurement of recreation-related norms should focus as directly as possible on impacts that are relevant to visitors. In this way, visitors are more likely to be able to report norms, norms are likely to be more highly crystallized, and management will be focused more directly on issues of concern to visitors. Data from several studies support the importance of this issue. For example, in the New River study noted earlier, a higher percentage of respondents reported a norm for waiting time to run rapids (while other boats took their turn) than for number of other boats seen (Roggenbuck et al. 1991). Similarly, visitors to the Clackamas River, Oregon, another relatively high-use area, reported norms more often for percentage of time in sight of other boats than for number of other boats seen (Hall et al. 1996). In relatively high-use areas, use levels may be perceived or manifested differently than in relatively low-use areas. Moreover, in high-use areas, it may simply not be feasible to estimate or evaluate large numbers of encounters with other groups. Several studies have explored alternative expressions of use-related indicators, including physical proximity of anglers along streams (Martinson and Shelby 1992), the number of people-at-one-time (PAOT) at destination or attraction sites (Manning et al. 1995a; Manning et al. 1995b; Manning and Lime 1996; Manning et al. 1996a; Manning et al. 1996b; Vaske et al. 1996; Manning et al. 1997), persons-per-viewscape (PPV) along trails (Manning et al. 1997a), and waiting times for essential services (Kim and Shelby 1998). In all of these cases, level of use is important in a generic sense, but how level of use is manifested to visitors as an impact that degrades the park experience varies by context. It is important that researchers and managers focus on impacts that are salient to visitors, especially when addressing the experiential component of carrying capacity.

Studies of park and recreation-related norms have used a variety of evaluative dimensions. When respondents are asked to evaluate a range of conditions for potential indicators, the response scale may use a variety of evaluative dimensions. The example of a social norm curve illustrated in figure 5.1 uses the evaluative dimension of *acceptability*, a commonly used response scale. However, other evaluative dimensions can and have been used, including *preference* (the condition that respondents prefer absent any other considerations), *displacement* (the point at which impacts are evaluated so negatively that respondents would choose not to visit the site again), and *management action* (the point at which impacts are evaluated so negatively that respondents would support restrictions on visitor use to ensure that conditions do not deteriorate further) (Manning et al. 1997a; Manning et al. 1999a). These alternative evaluative dimensions have substantially different meanings and usually result in significantly different norms. Several studies have included measures of both preferred (or "ideal") conditions and acceptable (or "maximum" or "tolerable") conditions (Young et al. 1991; Hammitt and Rutlin 1995; Watson 1995; Manning et al. 1997; Manning et al. 1999a). In all cases, preferred conditions for encounter-related variables are substantially lower—less than half—than acceptable conditions. The literature on norm theory described at the beginning of this chapter has suggested that norm measurement questions adopt more explicitly normative concepts and terminology, including the condition that managers "should" maintain (Heywood 1996). An initial test of this concept found that it yielded significantly higher encounter-related norms than the evaluative dimension of acceptability (Manning et al. 1997a; Manning et al. 1999a). None of the evaluative dimensions described earlier may be more "valid" than any others, but researchers and managers should be conscious of this issue and exercise appropriate care and caution in interpreting and applying study findings. For example, standards based on preference-related norms may result in very high-quality recreation experiences but may restrict access to a relatively low number of visitors. In contrast, standards based on acceptability or displacement may result in recreation experiences that are of only marginal quality but that allow access to a larger number of visitors. Studies that employ multiple evaluative dimensions may result in findings that enrich the information base on which standards might be formulated.

Studies of park and recreation-related norms have also used alternative question-and-response formats. Early in this chapter, it was noted that norms are sometimes measured using a long format whereby respondents are asked to evaluate a range of conditions of indicator variables. A short version of this question format has also been employed whereby respondents are asked to specify the maximum acceptable level of impact. Only one study has used both question formats, and this found that the short-question format yielded a slightly lower encounter-related norm (Manning et al. 1997a; Manning et al. 1999a).

Several studies have explored the range of response options that might be included in norm measurement questions (Roggenbuck et al. 1991; Hall et al. 1996; Hall and Shelby 1996; Donnelly et al. 2000). In particular, these studies have addressed the issue of whether respondents should be presented with an option that the indicator is important to them but that they cannot specify a maximum amount of impact that is acceptable. The principal argument in favor of this option suggests that respondents should not be "forced" into reporting a norm in which they have little confidence. The principal arguments against this option are that it may simply present some respondents a convenient way to avoid a potentially difficult (yet relevant) question, and when respondents select this option, it yields little information that is useful in informing a standard. Empirical tests directed at this issue found that respondents who chose this option were more like respondents who reported a norm (with respect to reactions to impacts and attitudes toward management) than those who reported that the indicator was not important to them (Hall and Shelby 1996). Moreover, use of this response option did not affect the value of the norm derived, though it did affect the variance or crystallization of the norm (Hall et al. 1996). These tests have also found that providing this response option generally lowers the percentage of respondents who report a normative standard (Donnelly et al. 2000). These findings suggest that when and where previous research has found that selected indicator variables are salient and/or that relatively large percentages of respondents have been able to report standards, a response option that the indicator under study is important, but that an associated standard cannot be reported, may not be warranted.

Crystallization of norms is an important research and management issue. As noted earlier in this chapter, crystallization refers to the level of agreement or consensus about norms. The more agreement about norms, the more confidence managers might have in using such data to formulate standards. Most norm-related studies have reported some measure of crystallization. Standard deviations of mean and median values of norms are used most frequently, but coefficients of variation and semi-interquartile ranges have also been recommended to allow comparisons across variables that use different response scales and to reduce the effects of extreme values (Roggenbuck et al. 1991; Hall and Shelby 1996). However, there are no statistical guidelines or rules of thumb to indicate what constitutes "high" or "low" levels of agreement or consensus, and there is some disagreement in the literature concerning how recreation-related norms might be interpreted. Ultimately, some degree of judgment must be rendered by managers. If there appears to be moderate to high levels of agreement over norms, then managers can incorporate study findings into their decisions with confidence. If there does not appear to be much agreement over norms, then managers might focus on resolving conflicts among visitors, consider zoning areas for alternative recreation experiences (and alternative associated standards), or formulate standards based on other considerations (e.g., legal and policy mandates, historic precedence, interest group politics).

A closely related issue to crystallization is how normative standards might vary with selected respondent characteristics, characteristics of those encountered, and situational variables. For example, a variety of norms have been found to be related to selected visitor characteristics, including organizational affiliation—activity groups versus environmental organizations (Shelby and Shindler 1992)—level of involvement with wilderness recreation (Young et al. 1991), country of origin (Vaske et al. 1995; Vaske et al. 1996; Budruk and Manning 2003), and ethnicity and race (Heywood 1993; Heywood and Engelke 1995; Stanfield et al. 2006). Research on the effect of characteristics of those encountered has focused primarily on type of activity. Encounter-related norms have been found to vary depending upon whether those encountered are fishers, canoers, or tubers (Vaske et al 1986); boaters or bank fishers (Martinson and Shelby 1992); or hikers or bikers (Manning et al. 1997a). Finally, norms have been found to vary in relation to a number of situational or locational variables, including along the river versus in campsites (Shelby 1981), type of recreation area (Shelby 1981; Vaske et al. 1986), use level (Hall and Shelby 1996; Lewis et al. 1996b; Shelby et al. 1988a), and periphery versus interior locations of parks (Martin et al. 1989).

As research on park- and recreation-related norms has matured, attention has focused on the issue of norm *congruence*, sometimes called *norm-impact compatibility* (Shelby and Vaske 1991). This issue concerns the extent to which respondents evaluate relevant aspects of the recreation experience in keeping with their normative standards. If recreation norms are to be used in formulating standards, then research on norm congruence is important to test the internal consistency or "validity" of such norms. A number of studies have addressed this issue across a variety of activities, indicator variables, and areas (Vaske et al. 1986; Patterson and Hammitt 1990; Hammitt and Patterson 1991; Williams et al. 1991; Ruddell and Gramann 1994; Hammitt and Rutlin 1995; Lewis et al. 1996b; Manning et al. 1996a; Manning et al. 1996e; Vaske et al. 1996). Nearly all have found support for the concept of norm congruence; that is, when conditions are experienced that violate visitor norms, respondents tend to judge such conditions as less acceptable (e.g., more crowded or degraded) and adopt behaviors to avoid such conditions (e.g., leave the park earlier than planned). Only one study has not supported norm congruence (Patterson and Hammitt 1990). However, this study was conducted in a relatively high-use area where encounter norms may not have been salient or highly crystallized and the study applied an especially strict test of congruence.

A variety of statistics are available for measuring, analyzing, and interpreting norms (Shelby and Heberlein 1986; Vaske et al. 1986; Whittaker and Shelby 1988; Shelby et al. 1996). Each has advantages and disadvantages, and these should be considered when selecting appropriate statistical approaches. Norms are generally reported and described in terms of means and medians. Mean values are intuitively straightforward and easy to calculate, but they can be skewed by outlying or extreme values and may be misleading in the case of multiple-tolerance norms. Median values have intuitive appeal because they represent the level of impact that half of respondents find acceptable. Social norm curves like the one illustrated in figure 5.1 (and elsewhere in this book), as well as frequency distributions like the one illustrated in figure 5.3 that show the level of agreement associated with each impact level, are less parsimonious but offer considerably more information in a graphic and less technical way. Statistical measures of norm crystallization were discussed earlier in this chapter.

Research has also examined the relationship between visitor-based norms and the conditions that are experienced in parks by visitors. This issue is potentially important because if visitors' normative standards are shaped primarily by existing park conditions (i.e., the current condition of park resources and experiences), then visitors might simply evaluate whatever sets of resource and social conditions they experience as acceptable (Roggenbuck et al. 1991; Stewart and Cole 2001; Stewart and Cole 2003). This could lead to a downward spiral in resource and/or social conditions. A test of this relationship was conducted using data collected from studies of visitor-based normative standards for fifty-six resource and social indicators in eleven national parks (Laven et al. 2005). In all of these studies, normative standards of visitors for selected resource and social indicators were measured, and this was followed by questions that asked respondents to report the condition of these indicator variables that was experienced in the park. Study findings suggest that there is generally little-to-no relationship between normative standards and existing conditions: of the 214 statistical tests (contingency coefficients) conducted, 128 (59.8%) found no statistically significant relationship. Of the remaining 86 statistically significant relationships, only 11 were classified as "strong." This suggests that normative standards of park visitors are generally derived independently of existing conditions.

The stability of park and recreation norms over time has received little research attention but may become increasingly important. Do personal and/or social norms change or evolve over time? If so, should such changes be incorporated into how parks and related areas are managed? The answer to the first question is a technical issue, while the second is more philosophical. Few studies have addressed the stability of norms over time. Those that have, generated mixed or inconclusive results. Studies covering the longest period of time (twenty-two years) were applied to the wilderness portion of Denali National Park, Alaska (Womble 1979; Bacon et al. 2001). Wilderness visitors in 2000 reported very similar normative standards for resource and social conditions to those reported by visitors in 1978. A 1977 study of encounter norms for boaters on the Rogue River, Oregon, was replicated in 1984 (Shelby et al. 1988a). No statistically significant difference was found for the number of acceptable river encounters. However, camp encounter norms were found to be significantly higher (or more tolerant) in the latter study. A similar study conducted in three wilderness areas over a longer interval found few clear, consistent trends in tolerance for intergroup contacts (Cole et al. 1995). Two other studies have found substantial stability of norms over time; however, these studies cover only a two- to three-year time period (Kim and Shelby 1998; Manning et al. 1999a).

Arguments about whether evolution of norms should be incorporated into management plans are divided. The underlying rationale of indicators and standards is that they should be set and maintained for some extended period of time, usually defined as the life of the management plan for which they are formulated. Thus, during this time period, standards probably should not be revised substantially. However, management plans are periodically reformulated to reflect the changing conditions of society. It seems reasonable to reassess park and recreation norms as part of this process and incorporate these findings into long-term planning processes. This approach is in keeping with contemporary notions of adaptive management, as discussed in chapter 19 (Stankey et al. 2005; K. Lee 1992).

Finally, as with all research methods, normative research must be tested for its *validity*. Validity is a complex issue and can be assessed in multiple ways (Carmines and Zeller 1979). For example, norm congruence, as described earlier in this chapter, constitutes a test of validity called *predictive* or *criterion* validity, and most studies have found that respondents evaluate park conditions and otherwise behave in ways that are consistent (or congruent) with their stated normative standards. Other approaches to assessing validity include *face* validity (e.g., the extent to which a research instrument "looks like" it measures what it is intended to measure, and the logic and consistency of study findings), and *construct* validity (the degree to which multiple variables that comprise a theoretical construct [in this case, normative standards] are represented in instruments designed to measure that construct). These approaches to validity have been assessed for normative research but primarily in the context of using visual research methods. These assessments of validity have been positive and are described more fully in chapter 6.

CHAPTER 6

Visual Research Methods

The National Parks . . . are open to all, without limit.

The normative research approach described in chapter 5 relies on an effective means of communication between researchers and respondents. For example, researchers may wish to present a range of visitor-caused impacts (e.g., increasing levels of use, increasing levels of trail impacts) to respondents for their evaluation, or respondents may simply be asked to report the minimum social and environmental conditions they find acceptable in parks and related areas. In many cases, this communication can be conducted in conventional numerical and/or narrative formats. For example, in parks where visitor-use levels are low, it may be reasonable to ask respondents to report the maximum number of other hiking groups per day it would be acceptable to see along trails. However, where use levels are relatively high, or when the impacts of visitor use are more complex and can be verbally described only in technical terms (e.g., level of trail erosion), visual approaches may be useful.

Visual research methods have played an important role in environmental science and natural resource management for many years (Daniel and Boster 1976; Ribe 1989; Shuttleworth 1980) and have been adapted for use in several dimensions of park and related research. For example, visual approaches have been used in studies assessing the aesthetic implications of forest harvesting and insect damage (Hollenhorst et al. 1993; McCool et al. 1986); the relative importance of campground attributes (Daniel et al. 1989); public perceptions of litter (Budruk and Manning 2004; Heywood and Murdock 2002); and recreation participation (T. Brown et al. 1989). The use of visual images, including slides and photographs, has been widely validated in the scientific literature (Bateson and Hui 1992; Daniel and Boster 1976; Daniel and Ittelson 1981; Daniel and Meitner

2001; Hershberger and Cass 1974; Hull and Stewart 1992; Kellomaki and Savo-lianen 1984; Stamps 1990).

The technology to develop visual representations of landscape and related social and managerial settings has improved dramatically in recent years. Computer aided design, geographic information systems, and computer animation can effectively be run from laptop computers rather than the mainframes that were once needed. Advances in digital cameras and photo-editing software provide both the resolution and editorial control that, in experienced hands, can result in edited images that are virtually indistinguishable from original photographs or slides. These widely accessible tools allow realistic and accurate depictions of potential settings and future conditions in a format that is familiar and easily understood.

Applying Visual Research Methods

Visual research methods offer a potentially important research approach that can be applied to measuring standards for parks and related areas, and they offer several potential advantages to narrative/numerical descriptions of certain park and outdoor recreation conditions (Manning and Freimund 2004). For example, visually based studies can provide additional pertinent information to respondents that would be difficult or awkward to communicate through conventional narrative/numerical approaches. For instance, in visual studies of crowding, all respondents see not only the same number of visitors encountered, but also potentially important characteristics of those encountered, including recreation activity engaged in, mode of travel, and group size. This is potentially important because perceived crowding has been found to be mediated by such variables (Manning 1986; Manning 1999; Manning et al. 2000). In more conventional narrative/numerical approaches, respondents may have to make assumptions about such characteristics and these assumptions are likely to vary among respondents. Visual research methods also focus directly and exclusively on the variables under study. For example, in visual studies of crowding, the number and type of visitors encountered is the only "treatment" allowed to vary, with all other variables held constant. Visual research methods can be especially useful in studying standards for indicator variables that are difficult or awkward to describe in narrative/numerical terms. For example, visual images of trail and campsite impacts may represent a more powerful and elegant means of communication with respondents than detailed and technical narrative descriptions. Finally, visual images can be edited to present conditions that are difficult to find in the field or that do not currently exist. For example, visual studies of crowding and resource impacts have incorporated images of conditions that do not now exist but will occur in the future as a function of continuing-use trends.

Visual images in the form of artistic renderings, photographs, computer-edited photographs, videotapes, and computer animations have been used to explore and assess visitor perceptions and evaluations of a range of park and outdoor recreation conditions, and this research approach has been increasingly applied in recent years to the issue of measuring and formulating park and related standards. For example, early studies used artistic renderings to represent a range of both resource and social impacts related to outdoor recreation. A study designed to explore potential differences in environmental perceptions between wilderness visitors and managers used a series of fourteen color drawings to illustrate a range of impact levels (Martin et al. 1989). A series of eleven pen and ink sketches of different types of visitor use was used to explore visitor norms and associated behavioral conventions in picnic areas (Heywood 1993). Photographs of wilderness campsites were used to measure normative standards for impacts to groundcover vegetation and fire rings (Shelby and Shindler 1992).

More recent research has used computer-edited photographs to measure visitor-based standards for selected components of parks and outdoor recreation areas and experiences. Sometimes called *Image Capture Technology (ICT)*, the representation and editing of photographic images using microcomputers has been used in a variety of settings to assess the visual quality of environmental conditions and to represent a spectrum of visitor use and impact conditions (Chenoweth 1990; Lime 1990; Nassauer 1990; Pitt 1990; Vining and Orland 1989). Scenic quality measurements made from photographic slides and the same images projected on a computer monitor have been found to be highly correlated (Vining and Orland 1989). No significant differences, for example, were found in aesthetic responses to color slides, computer monitor images, and images projected from videotape (Pitt et al. 1993). Recent work points out that representative validity is strongest when computerized visualizations are close in quality and definition to that of photographic slides (Daniel and Meitner 2001). The quality of ICT rendering is now often indistinguishable from original slides or photographs. These studies support extending ICT techniques to measurement of standards in parks and outdoor recreation, and this research approach has been increasingly applied in a variety of park and related contexts.

The study at Arches National Park, Utah, described in chapter 4, found that the number of visitors at attraction sites such as Delicate Arch was important in determining the quality of the recreation experience (Manning et al. 1995a; Manning et al. 1995b; Manning et al. 1993). A second phase of research was designed to measure visitor-based standards for the maximum acceptable number of visitors at such sites (Hof et al. 1994; Manning et al. 1996a; Manning et al. 1996b). A series of sixteen computer-edited photographs was prepared showing a range of visitors at Delicate Arch. Representative photographs are shown in figure 6.1. (This research was applied to other park sites as well.) The number of visitors in the sixteen study photographs ranged from 0 to 108, with the upper

(a) 0 people

(b) 12 people

(c) 52 people

(d) 108 people

Figure 6.1. Representative photographs of Delicate Arch showing a range of visitor use levels.

end of the range designed to show approximately 30% more visitors than the current maximum. The purpose was to illustrate a full range of density conditions, including the near-term future. A representative sample of visitors who had just completed their hike to Delicate Arch was asked to examine the photographs in random order and rate the acceptability of each on a scale that ranged from –4 ("very unacceptable") to +4 ("very acceptable"), with a neutral point of 0. Respondents were also asked to select the photograph that was most representative of the scene when they visited Delicate Arch and to report their degree of perceived crowding. Individual acceptability ratings were aggregated into a social norm curve (as described in chapter 5) and this provided an empirical foundation for helping to formulate a density-related standard for this site. The social norm curve for crowding at Delicate Arch (based on regression analysis of resulting data) is shown in figure 6.2.

Visual research methods have been expanded to address other social, resource, and managerial components of park and outdoor recreation areas and experiences. For example, outdoor recreation research suggests that perceived crowding may be influenced by visitor behavior, including recreation activities,

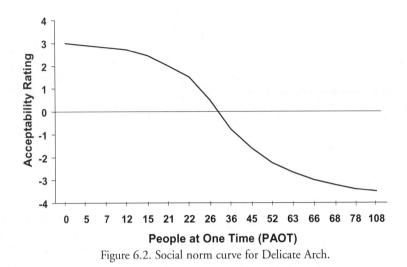

Figure 6.2. Social norm curve for Delicate Arch.

as well as density of use (Manning 1986; Manning 1999). Visual research methods have been used to assess the influence of visitor behavior on crowding-related standards. A study of crowding on the carriage roads (a multiple-use trail system) of Acadia National Park, Maine, used a series of nineteen photographs illustrating a range of use levels as well as alternative mixes of hikers and bikers, the two principal user groups (Manning et al. 1999a; Manning et al. 2000). Representative photographs are shown in figure 6.3. Study findings estimated crowding-related standards for the carriage roads and the influence of type of user group on such standards. Alternative crowding-related standards were found depending on the mix of recreation activities.

As noted, visual research methods have also been applied to selected resource-related impacts of outdoor recreation (Manning et al. 2004a; Martin et al. 1989; Shelby and Shindler 1992). For example, ecological research suggests that one of the principal impacts of recreation in wilderness is degradation of campsites through destruction of groundcover vegetation, soil compaction and erosion, injury to trees, and construction of multiple fire rings (Hammitt and Cole 1998; Leung and Marion 2000). To measure visitor-based standards for these impacts, a series of five computer-edited photographs was prepared illustrating a range of impacts to campsites in the wilderness portion of Yosemite National Park, California. Study photographs are shown in figure 6.4. The photographs were constructed on the basis of data from the park's long-term Wilderness Impact Monitoring System (Boyers et al. 1999). As part of a larger survey, a representative sample of wilderness visitors was asked to indicate which photograph was most like the campsite conditions they preferred to find, and a frequency distribution of preferred campsite conditions is shown in figure 6.5. A strong plurality of

(a) 0 people   (b) 5 people

(c) 10 people   (d) 15 people

(e) 20 people   (f) 30 people

Figure 6.3. Representative photographs of the Carriage Roads at Acadia National Park
showing a range of two types of visitor use (hikers and bikers).

respondents (40.8%) preferred to see no more impact than that represented in
study photograph 6.4(a), and a large majority of respondents (67.8%) preferred
to see no more impact than that represented in study photograph 6.4(b). These
data provide an empirical basis for helping to formulate standards for resource
conditions (at least their aesthetic dimensions) at this park.

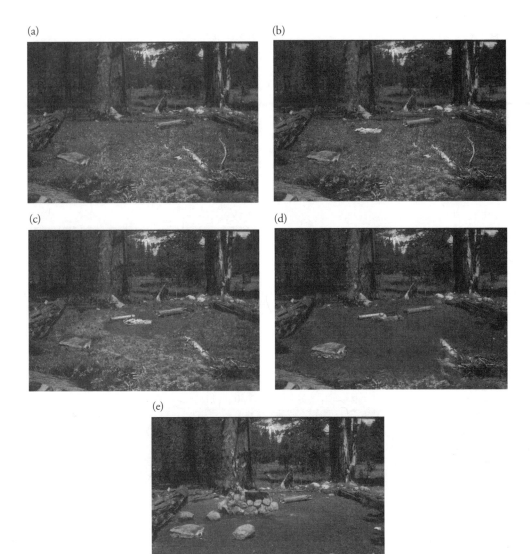

Figure 6.4. Study photographs illustrating a range of campsite impacts
in the wilderness of Yosemite National Park.

Visual research methods have also been applied to the managerial component
of parks and outdoor recreation areas. For example, recreation-caused impacts
to wilderness campsites (as described earlier) might be controlled through site
management practices, including defining the perimeter of the campsite (to dis-
courage campsite expansion), providing a hardened tent platform, and con-

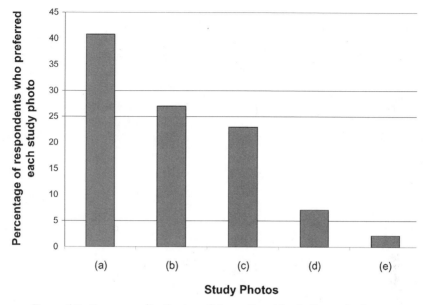

Figure 6.5. Frequency distribution of the preferred level of campsite impacts in the wilderness of Yosemite National Park.

structing a fixed fire ring. However, visitors may have normative standards concerning the type and level of such management practices and associated development. This issue was explored at Zion National Park, Utah, by developing a series of four computer-edited photographs illustrating a range of campsite management practices, as shown in figure 6.6. A representative sample of visitors who had received a wilderness permit were asked to rate the acceptability of each study photograph, and the resulting social norm curve is shown in figure 6.7. In this case, the minimal management approach represented by study photograph 6.6(a) is preferred, but the more aggressive management approaches represented by study photographs 6.6(b) and 6.6(c) are also judged acceptable. Only the management approach using nonnatural materials for site delineation as represented by study photograph 6.6(d) is judged as unacceptable. These data provide an empirical basis for formulating standards for management-related indicators.

Technological innovations in visual research methods continue to expand, including digital photography, desktop digital-editing software, and development of videotapes, compact disks (CDs), digital video disks (DVDs), and computer animation. Moreover, adoption of home computers and Internet access is also growing. These trends suggest an increasing variety of visual-based media that might be adopted in visual research methods designed to measure standards in parks and related areas. For example, a recent study incorporated

Figure 6.6. Study photographs illustrating a range of campsite development at Zion National Park.

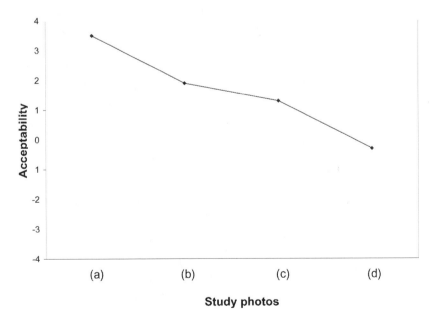

Figure 6.7. Social norm curve for campsite development at Zion National Park.

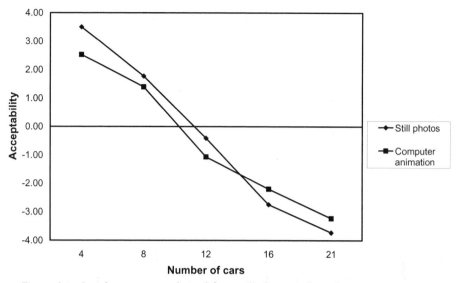

Figure 6.8. Social norm curves derived from still photographs and computer animation of a range of automobile traffic levels in Acadia National Park.

computer-edited images of a range of resource, social, and managerial conditions at Gwaii Haanas National Park Reserve, British Columbia, Canada, onto a videotape that was sent to a representative sample of park visitors (Freimund et al. 2002). The videotape included survey instructions. More than 75% of respondents reported that the images on the videotape served as useful reminders of their visit and helped them articulate their standards for recreation-related impacts.

Computer animation has also been used to represent the dynamic character of recreation use (Valliere et al. 2006). Initial research on standards for automobile traffic levels on scenic roads in Acadia National Park, Maine, used a series of five still photographs illustrating a range of traffic density on a representative section of the Park Loop Road. Subsequently, a series of five computer animations were developed to represent these same five density conditions on the same road section, albeit in a more dynamic and realistic manner. Each computer animation was an approximately thirty-second video clip. In each study, representative samples of park visitors were asked to view the photographs or video clips and rate the acceptability of each. Resulting social norm curves for both studies are shown in figure 6.8, and findings are generally comparable. This suggests that still photographs may be satisfactory in measuring standards for at least some indicator variables that are inherently dynamic.

Theoretical and Methodological Issues

Use of visual research methods in measuring standards in parks and outdoor recreation has raised a number of theoretical and methodological issues. These issues include the contexts in which visual research methods may be most appropriate, comparison of standards derived from visual research methods and more conventional narrative/numerical methods, validity of visual research methods, and methodological issues in applying visual research.

Application to Frontcountry and Other High-Density Contexts

Much of the research on crowding-related standards in outdoor recreation has focused on wilderness or backcountry areas. By definition, use levels in these areas are relatively low. In this context, a narrative/numerical approach to measuring standards is probably appropriate. Standards-related questions usually take one of two such forms: respondents are asked to rate the acceptability of encountering various numbers of other visitors or groups of visitors, or respondents are simply asked to report the maximum acceptable number of encounters.

However, in frontcountry and other relatively high-use contexts, this measurement approach may be less appropriate. In such high-use areas it may be unrealistic to expect respondents to accurately judge or report the maximum acceptable number of visitors or groups of visitors. The research literature is suggestive of this issue. First, several studies have found that respondents are less likely to be able to report a discrete maximum acceptable number of encounters in relatively high-use areas as compared to relatively low-use areas (Roggenbuck et al. 1991; Shelby and Vaske 1991; Vaske et al. 1986). Second, there tends to be less consensus about such crowding-related standards in relatively high-use areas, and this may be due at least in part to measurement error (Manning 1999). Third, there is evidence that self-reports of encounters by visitors in relatively high-use areas are not accurate. A study of river use found that floaters who experience fewer than six encounters per day with other river users generally were able to report them accurately (by comparison with actual encounters as counted by a trained observer) (Shelby and Colvin 1982). But at higher levels of use, most visitors underreported encounter levels.

Thus, in frontcountry or other high-use density contexts, visual research methods may be more appropriate than conventional narrative/numerical methods, because they do not require visitors to accurately keep track of and report discrete numbers of other visitors or groups of visitors encountered (or that are acceptable). For this and other reasons (as outlined in the following section), visual research methods may offer more valid estimates of crowding-related standards, especially in high-density settings.

However, it should be noted that in some high-use areas, the absolute number of other visitors encountered (along walkways or at attraction sites) may not be an especially salient indicator variable. As suggested in chapter 5, crowding can be manifested in potentially many ways, including waiting times to access visitor attractions (Budruk and Manning 2003). In such cases, narrative/numerical question formats to elicit visitor-based standards may be appropriate and effective.

Comparison of Visual and Narrative/Numerical Research Methods

A related issue concerns comparison of crowding-related standards derived from visual and narrative/numerical research methods. A test of this relationship was conducted as part of the research at Arches National Park, Utah, described earlier (Manning et al. 1996a). A social norm curve derived from respondent ratings of the acceptability of the sixteen photographs illustrating a range of visitors at Delicate Arch estimated a crowding-related standard of approximately twenty-eight visitors at one time (the point at which aggregate acceptability ratings fell out of the acceptable range and into the unacceptable range). The social norm curve is shown in figure 6.2. Using a narrative/numerical approach, respondents were also asked to report a discrete maximum number of visitors at one time acceptable at the arch. The average number of visitors reported was just under seventeen, suggesting a substantially lower crowding-related standard than derived from the visual research method.

The crowding literature may help to explain why crowding standards derived from visual research methods are substantially higher than those derived from more conventional narrative/numerical methods, and why the former may be a more valid or realistic estimate. Studies of crowding in outdoor recreation indicate that perceived crowding may be a function of several categories of variables, including the characteristics of respondents, the characteristics of visitors encountered, and situational or environmental variables (Manning 1986; Manning 1999; Manning et al. 2000). The second category of variables—the characteristics of visitors encountered—may be of particular interest when comparing visual and narrative/numerical research methods. There is considerable evidence in the literature that the characteristics of visitors encountered can affect crowding-related standards. Factors found important include the type and size of the group, visitor behavior, and the degree to which groups are perceived to be alike. For example, several studies have found differential crowding effects based on nonmotorized versus motorized boats (Lucas 1964), hikers and horseback riders (Stankey 1973; Stankey 1980a), and small versus large groups (Lime 1972; Stankey 1973). In all of these cases, encounters with one type of visitor (the latter type in the above cases) has greater impact on perceived crowding than encounters with the other type of visitor.

Similarly, inappropriate behavior (e.g., noncompliance with rules and regulations, boisterous behavior) can contribute in important ways to perceived crowding. In fact, several studies indicate that such behavior can have a greater impact on perceived crowding than sheer numbers of encounters (Driver and Bassett 1975; Titre and Mills 1982; West 1982).

Finally, perceived alikeness between groups can affect judgments about crowding. This concept might best be understood through appreciation of the role of social groups in outdoor recreation. Numerous studies have emphasized the importance of the social group in outdoor recreation: the vast majority of people participate in outdoor recreation in family and friendship groups (Buchanan et al. 1981; Burch 1964; Burch 1969; Cheek 1971; Dottavio et al. 1980; Field and O'Leary 1973; Meyersohn 1969). This suggests the notion of solitude so often associated with certain types of outdoor recreation may not mean simple isolation from others. It also suggests an inward focus on interpersonal relationships within the social group. Several studies have developed empirical insights that begin to link the concepts of social groups, solitude, perceived alikeness between groups, and crowding-related standards. For example, it has been demonstrated that solitude is a multidimensional concept and that, in the context of outdoor recreation, solitude may have more to do with interaction among group members free from outside disruptions than with physical isolation (Twight et al. 1981; Hammitt 1982). This suggests that as long as encounters with other groups are not considered to be disturbing, they do not engender strong feelings of crowding. And this, in turn, suggests the notion of perceived alikeness. In particular, it has been suggested that much of the social interaction between groups in outdoor recreation settings is conducted with little conscious deliberation, or, in more technical terms, in nonsymbolic modes of communication (R. Lee 1972; R. Lee 1975; R. Lee 1977). Such communication has been defined as "spontaneous and direct responses to the gestures of the other individual, without the intermediation of any interpretation" (Blumer 1936). People are therefore largely unaware of such social interaction, and it has little effect on perceptions of crowding. It can be concluded that the quality of a recreation experience "appears to be closely linked with the opportunity to take for granted the behavior of other visitors," and that "an essential ingredient for such an experience [is] the assumption that other visitors are very much like oneself, and will, therefore, behave in a similar manner" (R. Lee 1977). Thus, to the extent that groups are perceived as alike and require little conscious attention, encounters may have less impact on perceived crowding than might otherwise be expected.

The studies and ideas described may suggest why crowding-related standards developed from the traditional narrative/numerical approach might be most appropriately interpreted as the lower bounds of acceptability. The crowding literature illustrates that all contacts do not contribute equally to perceived crowding. However, studies that query respondents directly about appropriate encounter levels (i.e., narrative/numerical studies) contain an implied assumption that all encounters are similar. Moreover, such studies by their very nature focus on encounters that require full and explicit attention by the respondent. In other words, they present the worst case. Encounters between groups that are similar and thus may require and receive little conscious attention, and may have relatively little effect on perceived crowding, are left unconsidered. Crowding-related standards based on narrative/numerical research methods might be increased to the extent that groups are compatible in mode of travel, size, behavior, and other factors that contribute to perceptions of alikeness.

Based on this reasoning, visual research methods may represent a more realistic approach to measuring crowding-related standards. Respondents are able to examine a visual portrayal of use conditions, including at least some relevant characteristics of those encountered (e.g., recreation activity, mode of travel, size of group). It is likely that some of the visitors portrayed in these scenes may not consciously register in the minds of respondents because they are perceived to be like the respondent in important ways. The differences in crowding-related standards found in studies comparing visual and narrative/numerical research methods tend to support this idea empirically.

Findings from the study of crowding-related standards on the carriage roads of Acadia National Park, described earlier, provide additional empirical support for the conceptual ideas discussed (Manning et al. 2000). In this study, a series of computer-edited images presented both a range of use levels along the carriage roads and alternative mixes of the primary user groups—hikers and bikers. The mixes of hikers and bikers ranged from equal distributions to exclusively either hikers or bikers. Study findings suggest that crowding-related standards are influenced by both the number and type of users. For example, when the social norm curve (derived from the mean acceptability ratings for each photograph) was constructed for the subpopulation of respondents who were hiking, the curve fell out of the acceptable range and into the unacceptable range (i.e., crossed the neutral or zero point on the acceptability scale) at sixteen visitors for the series of photographs that showed a range of exclusively hikers. However, the social norm curve for the same subpopulation of respondents who were hiking crossed the neutral point on the acceptability scale at ten visitors for the series of photographs that showed the same range of exclusively bikers. These findings support the notion that crowding-related standards can be influenced in a substantive way by presenting information on type of recreation activity. This type of information can be presented effectively and subtly through visual methods but may be too complex and explicit to be effectively presented in a conventional narrative/numeric manner.

Validity of Visual Research Methods

As visual research methods (and normative research approaches more generally) are increasingly applied to measure standards in parks and outdoor recreation, it is important that the validity of this research be assessed. However, as suggested in chapter 5, the issue of validity is complex and can be assessed in multiple ways (Carmines and Zeller 1979; Nunnally 1978). In its most generic sense, the concept of validity refers to the degree to which a research instrument does what it is intended to do, or measures what it purports to measure. To what degree do visual research methods for measuring standards provide valid estimates of the minimum acceptable conditions of parks and related areas? Several approaches to measuring validity may be appropriate to answering this question.

Face validity is a conventional approach to assessing validity and refers to the extent to which an instrument "looks like" it measures what it is intended to measure. Studies incorporating visual research methods in measuring standards for parks and outdoor recreation might contribute to assessing face validity in two ways. First, several studies have adapted and applied a "verbal protocol assessment" (Schkade and Payne 1994) designed to assist respondents in considering the degree to which they understood study questions and the extent to which they are confident in their answers (Manning et al. 2001). In these studies, a series of statements was presented to respondents at the conclusion of visitor surveys employing visual research methods to measure normative standards, and respondents were asked to indicate the extent to which they agreed or disagreed with these statements:

1. I understood the questions that were asked.
2. The photographs realistically represent different levels of use at this area.
3. I was confused by the questions that asked me to choose between the photographs.
4. It was very difficult to rate the acceptability of the photographs.
5. The answers I gave to these questions accurately represent my feelings about acceptable use levels on the trails I hiked.
6. The National Park Service [NPS] should manage visitor use levels based on the kind of information collected in studies like these.

The verbal protocol assessment was administered in conjunction with visually based visitor surveys administered at several sites at Grand Canyon National Park, Arizona, Arches National Park, Utah, and Yosemite National Park, California.

Nearly all respondents at all three parks agreed that they understood the questions that were asked. Similarly, the vast majority of respondents agreed that the photographs used in the studies realistically represented different levels of use at the study sites. A majority or plurality of respondents reported that they were not confused by the questions that asked them to choose the photograph that represented the highest acceptable level of use, and that it was not difficult to rate the acceptability of the photographs. The vast majority of respondents agreed that their answers to the crowding-related questions accurately represented their feelings about acceptable use levels at the study sites. Finally, a strong majority of respondents agreed that the NPS should manage visitor-use levels based on the type of information collected in these kinds of studies. Similar findings were reported in a study at Gwaii Haanas National Park Reserve, British Columbia, Canada (described earlier), where 80% of the respondents agreed that the information gained on the videotapes "was a worthwhile addition to the paper questionnaire" (Freimund et al. 2002).

A second way of assessing face validity concerns the logic and consistency of study findings derived from visual research approaches. Three approaches might be used to explore this issue. First, the social norm curve shown in figure 6.9 is a representative example of those derived from visual research methods. Data used to construct the figure are from a study employing six computer-edited photographs illustrating a range of use levels in a strategic location in the prison cell house of Alcatraz Island, a unit of Golden Gate National Recreation Area, California. This study is described in more detail in chapter 13 (Manning et al. 2002a). The points defining the social norm curve are mean acceptability ratings for the six photographs. As would be expected, average acceptability ratings declined with increasing use levels, and there is a strong statistical relationship between these variables, with the number of visitors in the photographs explaining 58% of the variance in acceptability scores.

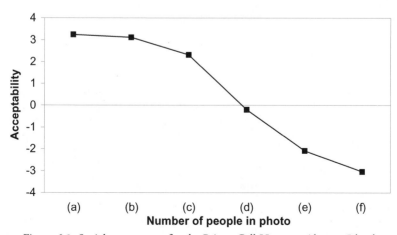

Figure 6.9. Social norm curve for the Prison Cell House at Alcatraz Island.

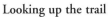

Figure 6.10. Representative photographs of the Bright Angel Trail presented from alternative landscape perspectives.

A second approach to examining the logic and consistency of study findings from visual research concerns the use of alternative *evaluative dimensions*. The studies at Grand Canyon, Arches, and Yosemite national parks incorporated four evaluative dimensions (described in chapter 5) in measuring visitor-based standards: preference (the condition respondents preferred), acceptability (the maximum level of impact respondents judged acceptable), management action (the maximum level of impact respondents felt the NPS should allow before limiting visitor use), and displacement (the level of impact that would keep respondents from visiting the park again) (Manning et al. 1999a; Manning 2001). Logic suggests that crowding-related standards estimated from these alternative evaluative dimensions would be ordered, with the lowest standards associated with the preference dimension, the highest standards associated with the displacement dimension, and the acceptability and management action-related standards near the midpoint of the range. This pattern of findings was consistent across all sample sites at all three study parks.

Another approach to assessing validity concerns the concept of *predictive* or *criterion* validity. This approach examines the correlation between findings derived from a study instrument and some important form of behavior that is external to the instrument, the latter referred to as the criterion. The concept of norm congruence (described in chapter 5) offers a test of criterion validity. Congruence refers to the extent to which visitors behave in relation to their stated standards (Manning 1999; Shelby and Heberlein 1986). Data from the study of visitors to Delicate Arch reported earlier in this chapter offers a test of congruence (Manning et al. 1996a). Three variables were used to test congruence: (1) the visitor-based standard for the maximum acceptable number of visitors at Delicate Arch, (2) the number of visitors in the photograph that respondents reported as best representing the density condition when they visited Delicate Arch, and (3) a measure of perceived crowding at Delicate Arch. It was hypothesized that if respondents experienced more visitors at Delicate Arch than the visitor-based standard, they would rate the experience as "crowded," at least to some degree. Likewise, if they experienced fewer visitors than the standard of quality, they would rate the experience as "not at all crowded." Study findings showed that the vast majority of respondents (74%) fell into one of these two categories of congruence.

A fourth conventional approach to assessing validity applies the concept of *construct* validity. This approach to validity examines the degree to which multiple variables that comprise a theoretical construct are represented in instruments designed to measure that construct. Measures of crowding-related standards are ultimately aimed at the theoretical construct of crowding. As noted earlier in this chapter, normative interpretation of crowding in outdoor recreation has generally recognized three broad types of variables as mediating perceived crowding: (1) characteristics of respondents (e.g., recreation activity in which the respondent is engaged); (2) characteristics of those encountered (e.g., recreation activity in which those encountered are engaged): and (3) situational variables (e.g., location in which encounters occur) (Manning 1986; Manning 1999). Visual research methods applied to measuring crowding have begun to incorporate all three types of these variables. For example, the study of carriage road use at Acadia National Park described earlier in this chapter used a visual research approach to measure crowding-related standards for two types of respondents/trail users (hikers and bikers), for encountering two types of trail users (hikers and bikers), and for two types of trails (high- and low-use trails) (Manning et al. 2000). In the study of Gwaii Hannas National Park Preserve described earlier, respondents discriminated clearly among the standards they held for encounters with kayakers or facilities as the context of the encounter changed. For example, evaluations of an encounter with nine other kayaks at one time at an "attraction site" were acceptable, but this level of encounters was unacceptable in a "wild place" (Freimund et al. 2002). Inclusion of multiple variables or dimensions of the theoretical construct of crowding into visually based measures of crowding-related standards can be seen to enhance the power and resolution of such measures as well as contributing to their construct validity.

The concept of validity is complex, and might most appropriately be described as an objective to which research should aspire rather than an end to be reached. In the words of Nunnally (1978, 87), "Validity is usually a matter of degree rather than an all or none property, and validation is an unending process." Validity can be assessed through theoretical, empirical, and common sense approaches. Findings from the studies described above tend to support the validity of visual research and the application of normative research methods to standards in parks and outdoor recreation.

Methodological Issues

As application of visual research methods to measuring park-related standards proceeds, methodological issues have arisen. For example, in other environmental applications of visual research methods, the landscape perspective of photographs may influence assessments of environmental conditions reported by respondents (T. Brown et al. 1989; Daniel and Boster 1976; Hollenhorst et al. 1993). This issue was explored in the context of measuring crowding-related standards in parks (Manning et al. 2002b). As part of the visitor survey at Grand Canyon National Park described earlier in this chapter, two sets of computer-edited photographs were prepared to illustrate a range of visitor-use levels on the Bright Angel Trail, the principal trail that connects the South Rim of the canyon with the Colorado River. Representative study photographs are shown in figure 6.10. Both sets of photographs showed the same range of visitor-use levels along the same fifty-meter section of trail. However, one set of photographs was prepared looking "up" the trail (showing a characteristically "closed in" view,) while the other set of photographs was prepared looking "down" the trail (showing a characteristically "open" view). Half the sample of hikers viewed the former set of photographs and half viewed the latter. Study data indicate virtually no differences in the crowding-related standards reported by respondents.

Starting point bias represents another potential methodological issue associated with visual research methods (as well as more conventional narrative/numerical methods). Research on willingness-to-pay for environmental amenities suggests that the initial monetary values presented to respondents may influence the ultimate value derived from the research (Desvousges et al. 1983; Rowe et al. 1980; Thayer 1981; Manning et al. 1999b). To explore this issue in the context of using visual research methods in measuring park and outdoor recreation standards, respondents at one site in the Grand Canyon National Park study described earlier were split into two subsamples. The first group of respondents was shown the six computer-edited study photographs of a range of visitor-use levels in increasing order of use density, while the other group of respondents saw the photographs in decreasing order. Study data indicate no substantive differences in the crowding-related standards reported by the two groups of respondents.

Finally, placement of individuals in study photographs may influence crowding-related standards. For example, in the study of Delicate Arch described earlier (and reported in figures 6.1 and 6.2), individuals in the foreground of study photographs were found to influence acceptability ratings to a greater degree than individuals in the background. Subsequent visually based research has been careful to distribute individuals equally in the foreground and background of study photographs.

While there are likely to be many methodological issues inherent in visual research methods as they are applied to measuring standards in parks and other outdoor recreation areas, initial research suggests that these methods may be relatively robust. That is, careful applications do not appear to be heavily influenced by methodological variations.

Assessing Visual Research

Visual research methods have played an important role in environmental research and management for several decades. More recently, these methods have been adapted for use in measuring normative standards in parks and related outdoor recreation areas. Study findings suggest that visual research methods may have some advantages over more conventional narrative/numerical research approaches, and that visual research methods may be particularly appropriate in selected park and outdoor recreation contexts, such as frontcountry and other high-use areas, and for resource-related impacts that are difficult to describe in narrative/numerical formats. Moreover, in certain contexts (e.g., high-use areas) visual research methods may result in more realistic estimates of visitor-based standards. Findings from studies employing visual research methods generally meet conventional tests of research validity. Finally, tests of selected methodological issues inherent in visual research approaches suggest that these methods may be relatively robust in that resulting data do not appear to be greatly influenced by methodological alternatives. Visual research methods are being increasingly adopted into studies of park and outdoor recreation standards and have received strong endorsements in the literature. For example, a recent analysis concluded that "the short phrases used in normative questions (such as 'number of encounters per day') cannot capture the true complexity and nature of a recreation experience and respondents must inevitably fill in background assumptions and conditions . . . Thus, we feel that . . . visual approaches are superior to the traditional . . . form of numerically based question[s]" (Hall and Roggenbuck 2002, 334).

Although visual research methods are promising as an approach to measuring standards in parks and outdoor recreation, there are several issues that warrant attention and that may limit their usefulness. A photograph can portray a more realistic description of a recreation setting than can a number or a short narrative statement, but there are limitations to what a photograph can present. It is unrealistic to expect that photographs can display all relevant characteristics of visitor use and users and associated impacts. Moreover, still photographs are (by definition) static, only account for visual stimuli, and may not be well suited to representing the inherent dynamics of a recreation experience. Video photography, computer animation, and other dynamic media, as well as audio recordings, may represent at least a partial solution to this issue. For example, a recent study has shown that nonvisual variables, such as sound and smell, can affect perceived crowding (Rohrmann and Bishop 2002). As noted earlier, computer animations have been used to help formulate standards for automobile traffic levels at Acadia National Park (Valliere et al. 2006) (though the normative standards derived were not substantively different from those derived from still photographs), and audio recordings of visitor-caused noise have been used to help formulate standards for "natural quiet" and soundscape management at Muir Woods National Monument, California (described in chapter 14) (Manning et al. 2006).

CHAPTER 7

Tradeoffs in Park and Outdoor Recreation Management

It is not mathematically possible to maximize for two (or more) variables at the same time.

Tradeoffs are an inherent and challenging element of park and outdoor recreation management. Indeed, in a fundamental way, tradeoffs are at the heart of the tragedy of the commons, carrying capacity, and related issues: to what extent and in what ways can we use the environment without spoiling what we value most about it? This question addresses the fundamental tradeoff between use and preservation. In its more specific manifestation as applied to national parks and protected areas, this issue has been codified in the two-fold mandate of the U.S. national parks, as described in chapter 2. How do we balance these competing objectives?

This generic question gives rise to a number of more specific manifestations. For example, in chapter 2 it was noted that the experience of visiting a park or related area comprises three basic dimensions: the resource conditions experienced (e.g., the amount of human impact at camping sites); the social conditions experienced (e.g., the number of other groups camped within sight or sound); and the management conditions imposed (e.g., the number of camping permits allowed) (Hendee and Dawson 2002; Manning 1999). In general, most visitors to national parks and related areas are thought to prefer a relatively pristine, natural environment, relatively few encounters with other groups, and a high degree of freedom from management control. While this is the ideal, attempts on the part of park managers to provide ideal conditions along one dimension of the park experience typically involve having to make concessions along one or both of the other dimensions of the park experience. As a result, decisions about how to manage parks involve inherent tradeoffs among the conditions of resource, social, and managerial attributes of the park experience. For example, the number of permits issued for recreational use of a park could be increased to allow more public access, but this might result in more resource impacts and encounters among groups within the park. Conversely, limiting the number of recreational-use permits issued might reduce resource impacts and encounters among groups but would allow fewer people to use and enjoy the park.

The issue of tradeoffs was introduced in chapter 5, where it was suggested that the "management action" dimension of normative evaluation is designed to address tradeoffs between the desire to protect the resource and social components of the park experience and the desire to maintain public access to parks. In normative questions employing the management action evaluative dimension, respondents are asked to report the minimum acceptable condition of the resource and social components of the park experience before restrictions on visitor use are imposed. Study findings demonstrate that management-action norms are significantly different (more tolerant) than preference-based norms, and that these differences are a function of the inherent tradeoffs in park and outdoor recreation management. Normative data derived from the context of such tradeoffs can be useful in informing park managers about carrying capacity-related indicators and standards, or the "mutual coercion, mutually agreed upon" desired by society.

While this contextual approach to normative research is useful, it can be supplemented by other research approaches to exploring tradeoffs in parks and outdoor recreation. The remaining two sections in this chapter describe *indifference curve analysis* and *stated choice modeling*, which have been adapted and applied to studying tradeoffs in park and outdoor recreation management in greater depth.

Indifference Curve Analysis

Developed in the discipline of economics, indifference curve analysis is designed to explore tradeoffs in consumer decision making. Indifference curve analysis provides a theoretical model representing the tradeoff decisions an individual makes in allocating a fixed level of income between two consumer goods (Nicholson 1995). There are two primary components to the indifference curve model: the individual's indifference curves, and his/her budget constraint. A single indifference curve represents all possible combinations of two goods (e.g., *A* and *B*) that provide an individual with the same level of utility (Pindyck and Rubinfeld 1995). The curves labeled IC_1 and IC_2 in figure 7.1 are examples of indifference curves. Indifference curves that include greater amounts of goods *A* and/or *B* represent a higher level of utility or satisfaction than those closer to the origin of the graph. The budget constraint represents the possible combinations

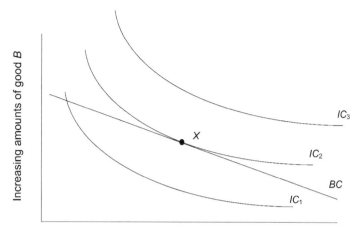

Figure 7.1. Theoretical indifference curves and budget constraint.

of goods A and B the individual can purchase, assuming the individual spends all of his/her income on the two goods (Pindyck and Rubinfeld 1995). For example, the budget constraint labeled BC in figure 7.1 represents all possible combinations of the two consumer goods A and B, for a fixed-income level.

According to indifference curve theory, the optimal combination of goods A and B for a given income is located where the budget constraint is tangent to the individual's highest indifference curve (Nicholson 1995). This represents the highest level of utility the individual can achieve from the consumption of goods A and B, given a fixed level of income. In figure 7.1, the optimal condition is represented by point X.

An adaptation and application of indifference curve analysis to park management was conducted within the context of carrying capacity at Arches National Park, Utah, by substituting solitude at Delicate Arch and access to Delicate Arch for consumer goods (i.e., goods A and B in figure 7.1) (Lawson and Manning 2000; Lawson and Manning 2001a; Lawson and Manning 2001b; Lawson and Manning 2002a). Specifically, the number of people at Delicate Arch was substituted for good B along the y axis, and the chance (%) of receiving a hypothetical permit to hike to Delicate Arch was substituted for good A along the x axis.

Indifference curves were estimated following a procedure in which respondents are presented with a series of pairs of solitude and access conditions (MacCrimmon and Toda 1969). The first component of each pair of conditions is a fixed reference point against which respondents evaluate a unique alternative condition. Respondents are asked to indicate their preference within each pair of conditions they evaluate. For example, in the study at Arches National Park, respondents were asked to express their preference between a first set of conditions—having a 100% chance of receiving a permit to hike to Delicate Arch and seeing 108 people at Delicate Arch, and a second set of conditions—having a 50% chance of receiving a permit to hike to Delicate Arch and seeing 36 people at the arch. (The use levels at Delicate Arch were represented using the photographs of visitor use at this site described in chapter 6 and illustrated in figure 6.1).

Regression analysis was used to estimate an indifference curve for each respondent based on the data points derived from the respondents' evaluation of a series of pairs of access and crowding conditions at Delicate Arch. For each respondent, a hyperbolic, semilog, or quadratic curve was fit to the data points. The functional form for each individual indifference curve was selected based on the goodness of fit (R^2) of the regression equation, and the explanatory significance of the access variable (chance of receiving a permit) on the number of people at Delicate Arch.

Analysis of sample data resulted in a total of sixteen unique indifference curves estimated. Respondents were categorized into one of three groups based on the slope and form of their indifference curves as illustrated in figure 7.2. (Readers may note that the y axis of this figure may appear to be labeled in an unintuitive manner. That is, the zero point appears at the top of the scale. This is because zero other people at Delicate Arch represents the highest degree of solitude. It will be remembered from figure 7.1 that the axes of indifference curves are constructed to show increasing "amounts" of the goods—or better conditions of such goods—under study.) The first group includes individuals whose preferences are "access oriented." The indifference curves for these individuals are characterized by steep slopes, meaning that "access oriented" respondents would tolerate large increases in the number of people seen at Delicate Arch to help ensure they would be granted access to this site. The second group includes individuals whose preferences are "solitude oriented." The indifference curves for these individuals are characterized by flat slopes up to the threshold, meaning that "solitude oriented" respondents would tolerate substantial reductions in their chances of receiving access to Delicate Arch to help ensure that, if they received access to the arch, they would see relatively few people. The third group includes individuals whose preferences are "tradeoff oriented." The indifference curves for these individuals are characterized by moderate slopes, meaning that "tradeoff oriented" respondents would prefer tradeoffs between solitude and access of a more proportional nature.

Based on the shape and slope of the indifference curves estimated, nearly half of all respondents (48.8%) revealed preferences characterized as "solitude oriented" compared with just one-fifth of respondents (20.3%) having preferences characterized as "access oriented." Just under one-third of respondents (30.9%) had preferences that were characterized as being "tradeoff oriented." The shape

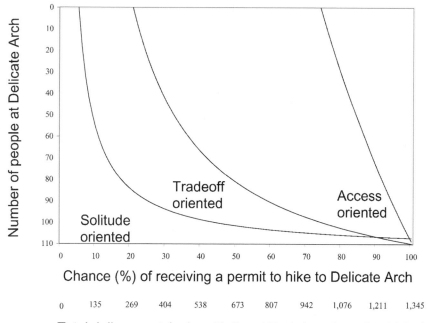

Figure 7.2. Indifference curves—access, solitude, and tradeoff oriented visitors.

and slope of these indifference curves "map" the tradeoffs visitors prefer to make between solitude and access at Delicate Arch and provide a more "contextually" informed basis for formulating crowding-related standards.

The budget constraint line for this analysis was constructed by developing a computer simulation model of visitor use at Arches National Park. (Computer simulation modeling of park use is described in chapter 8.) The model estimates the number of visitors at Delicate Arch based on the number of vehicles that enter the park each day. The model was run at a range of daily park-use levels to derive the possible combinations of the number of visitors at Delicate Arch and the probability of receiving a permit to hike to the arch. Adding the budget constraint line to the indifference curves allows estimation of the point at which utility is maximized or carrying capacity has been reached.

Stated Choice Modeling

Indifference curve analysis is useful in conceptualizing and measuring tradeoffs between two competing values, such as solitude and access, and thereby informing carrying capacity analysis. However, its application is limited to bivariate contexts. As noted earlier, the literature in outdoor recreation suggests that park experiences comprise three dimensions: resource, social, and managerial conditions (Manning 1999). Moreover, there can be multiple indicators of quality within each of these dimensions. Unfortunately, not all of these indicators can be optimized simultaneously, and tradeoffs must often be made among them, especially when demand for parks is high.

Stated choice modeling has been developed as a survey and related statistical technique to explore tradeoffs among multiple attributes of a good or service and is often applied in several types of consumer research (Green and Srinivasan 1978). In stated choice modeling, respondents are asked to make choices among alternative configurations of a multi-attribute good (Louviere and Timmermans 1990). Each alternative configuration is called a profile and is defined by varying levels of selected attributes of the good (Mackenzie 1993). For example, in the context of parks and outdoor recreation, respondents might be asked to choose between alternative park settings that vary in the number of other groups encountered, the quality of the natural environment, and the intensity of management regulations imposed on visitors. Respondents' choices among the alternatives are evaluated to estimate the relative importance of each attribute to the overall utility derived from the recreational setting. Further, stated choice analysis models are used to estimate public preferences or support for alternative combinations of the attribute levels (Dennis 1998).

Stated choice modeling was applied to study wilderness use in Denali National Park, Alaska (Lawson and Manning 2001c; Lawson and Manning 2002b; Lawson and Manning 2002c; Lawson and Manning 2002d). Based on review of the wilderness recreation literature and consultation with park staff, six wilderness setting attributes (or indicators) were selected to define the resource, social, and management conditions in the wilderness portion of the park. Each of these indicators was further defined by a range of three levels (or standards). Indicators and standards used in the study are shown in table 7.1. Given three standards for each of the six study indicators, a full factorial design would produce a total of 3^6 (729) hypothetical Denali wilderness settings. This number of alternatives is far too large for survey respondents to reasonably consider. Therefore, an orthogonal fractional factorial design was constructed containing nine pairwise comparisons (Seiden 1954). An example of a representative Denali wilderness setting comparison is presented in figure 7.3. Questionnaires were administered to a representative sample of overnight visitors to the wilderness portion of Denali. In each of the nine choice questions included in each version of the questionnaire respondents were asked to read each of the two wilderness-setting descriptions (A and B) and indicate which they preferred.

The responses to the stated choice questions were analyzed using logistic regression analysis to estimate a linear utility-difference model (Hosmer and Lemeshow 2000; Opaluch et al. 1993). The dependent variable of the model has

TABLE 7.1. Indicators and standards used
in wilderness study at Denali National Park, AK

Resource conditions
    Extent and character of hiking trails
    Hiking is along intermittent, animal like trails
        Hiking is along continuous single track trails developed from prior human use
        Hiking is along continuous trails with multiple tracks developed from prior
        human use
    Signs of human use at camping sites
        Camping sites have little or no signs of human use
        Camping sites have some signs of human use—light vegetation damage, a few
        moved rocks
        Camping sites have extensive signs of human use, e.g., bare soil, many rocks
        moved for wind protection and cooking
Social conditions
    Number of other groups encountered/ day while hiking
        Encounter 0 other groups per day while hiking
        Encounter 2 other groups per day while hiking
        Encounter 4 other groups per day while hiking
    Opportunity to camp out of sight and sound of other groups
        Able to camp out of sight and sound of other groups all nights
        Able to camp out of sight and sound of other groups most nights
        Able to camp out of sight and sound of other groups a minority of nights
Management conditions
    Regulation of camping
        Allowed to camp in any zone on any night
        Required to camp in specified zones
        Required to camp in designated sites
    Chance of receiving an overnight backcountry permit
        Most visitors are able to get a permit for their preferred trip
        Most visitors are able to get a permit for at least their second choice trip
        Only a minority of visitors are able to get a backcountry permit

an empirical and a theoretical interpretation. From an empirical perspective, the dependent variable is the log odds of choosing a given wilderness setting as the preferred alternative. In the context of random-utility theory, the underlying theoretical framework for this analysis, the dependent variable is interpreted as the ordinal utility or relative preference associated with a given wilderness setting. The independent variables of the model are a function of the six wilderness-setting indicators, which vary among the standards presented in table 7.1. Specifically, for each paired comparison question, the values of the independent variables are calculated as the difference in the levels of the six wilderness-setting indicators across the two wilderness-setting alternatives. The coefficients of the

Wilderness setting A	**Wilderness setting B**
• Encounter up to two other groups per day while hiking.	• Encounter up to four other groups per day while hiking.
• Able to camp out of sight and sound of other groups *all* nights.	• Able to camp out of sight and sound of other groups *most* nights.
• Hiking is along continuous, *single track* trails developed from prior human use.	• Hiking is along intermittent, animal-like trails.
• Camping sites have *some* signs of human use – light vegetation damage, a few moved rocks.	• Camping sites have *some* signs of human use – light vegetation damage, a few moved rocks.
• Required to camp at *designated sites*.	• Required to camp at *designated sites*.
• Only a minority of visitors are able to get a backcountry permit.	• Most visitors are able to get a backcountry permit for their *preferred* trip.

Figure 7.3. Example Denali wilderness setting paired comparison.

model provide estimates of the relative importance of the corresponding standards of the indicators to respondents. The coefficients of the model, together with their standard errors, *Wald Chi-Square* values, and *P* values are presented in table 7.2. All coefficients are significantly different than zero at the < .001% level, except the coefficients on "up to 2 other groups" and "intermittent animal like trails." The overall fit of the model is supported by the results of the Hosmer and Lemeshow goodness of fit test (x^2= 3.49, p = 0.836).

The magnitude of significant coefficients reflects the relative importance of the corresponding standard of the indicator to Denali overnight wilderness visitors. The values of the coefficients in table 7.2 imply that signs of human use at campsites influence Denali overnight wilderness visitors' utility or satisfaction more than any other wilderness-setting indicator considered in this study. Specifically, campsite conditions characterized as having "extensive signs of human use" are evaluated less favorably by Denali overnight wilderness visitors than any other standard of the six wilderness-setting attributes studied. Additionally, campsite conditions characterized by "little or no signs of human use" are pre-

TABLE 7.2. Coefficient estimates for wilderness-setting indicators at Denali National Park, AK

Indicators/Standards	Coefficient	Standard error	Wald Chi-Square	P Value
ENCOUNTERS WITH OTHER GROUPS/DAY WHILE HIKING				
0 other groups	0.440[a]	—	—	—
Up to 2 other groups	0.065	0.043	2.246	0.134
Up to 4 other groups	−0.504	0.044	132.826	*** 0.001
ABLE TO CAMP OUT OF SIGHT AND SOUND OF OTHER GROUPS				
All nights	0.295[a]	—	—	—
Most nights	0.145	0.044	11.148	*** 0.001
A minority of nights	−0.440	0.045	94.814	*** 0.001
HIKING IS ALONG				
Intermittent, animal-like trails	0.319[a]	—	—	—
Single track trails developed from human use	−0.028	0.044	0.403	0.526
Multiple-track trails developed from human use	−0.291	0.043	46.340	*** 0.001
CAMPING SITES HAVE				
Little or no signs of human use	0.582[a]	—	—	—
Some signs of human use	0.207	0.044	22.151	*** 0.001
Extensive signs of human use	−0.790	0.049	264.972	*** 0.001
REGULATION OF CAMPING				
Allowed to camp in any zone on any night	0.072	—	—	—
Required to camp in specified zones	0.140	0.048	8.620	0.003
Required to camp in designated sites	−0.212	0.045	21.948	*** 0.001
CHANCE VISITORS HAVE OF RECEIVING A PERMIT				
Most get a permit for their preferred trip	0.073[a]	—	—	—
Most get a permit for at least their second choice	0.143	0.044	10.424	***0.001
Only a minority get a permit	−0.216	0.043	24.656	*** 0.001

[a]Coefficients for the excluded standard of the indicator were not estimated by the statistical model. They were calculated as the negative sum of the coefficients on the other two standards of the corresponding indicator.

ferred more than any standard of any other wilderness-setting indicator included in the study.

The magnitude of the coefficient estimates in table 7.2 suggest that solitude-related indicators represent a second tier of importance to Denali overnight wilderness visitors. That is, while the number of encounters with other groups per day while hiking and opportunities to camp out of sight and sound of other groups are less important wilderness-setting indicators than campsite impacts, they demonstrate a relatively large influence on Denali overnight wilderness visitors' utility. The extent and character of trails, regulations concerning where visitors are allowed to camp in the Denali wilderness, and the availability of backcountry permits are less important to Denali overnight wilderness visitors relative to campsite impacts and solitude-related indicators of the Denali wilderness.

The relationship between the standards of each wilderness-setting indicator and the average utility associated with all possible combinations of the six Denali wilderness-setting indicators are plotted in figures 7.4(a)–(f). The values on the x-axis of each plot represent the standard of the corresponding Denali wilderness-setting indicator, and the values on the y-axis represent the amount by which the utility of the corresponding standard of the indicator deviates from average utility or satisfaction. The values on the y-axis are expressed in units of utility, which is a measure of relative preference. Standards of indicators with high utility values are preferred to standards of indicators with lower utility values. The plots provide further graphic insight into the relative importance of the wilderness-setting indicators to Denali overnight wilderness visitors. For example, utility drops sharply as campsites change from having "some signs of human use" (+0.207) to "extensive signs of human use" (–0.790) (figure 7.4[d]), whereas the loss of utility is less dramatic as the opportunity to camp out of sight and sound of other groups changes from "all nights" (0.295) to "most nights" (0.145) (figure 7.4[b]). In addition to measuring the relative importance of the six indicators included in the study, the graphs in figure 7.4(a)–(f) are also suggestive of potential standards by illustrating inflection or threshold points at which utility drops steeply or the point at which utility falls out of the positive range and into the negative range.

Stated choice modeling has also been applied to the wilderness portion of Yosemite National Park, California (Newman et al. 2005). Study data were used to develop a *management scenario calculator* that can create and test the acceptability of all permutations of indicators and standards included in the study. The calculator was created in Excel and is illustrated in figure 7.5. Like the Denali study described earlier, the Yosemite study included six indicators of wilderness conditions—two indicators of resource conditions (campsite impacts and impacts caused by stock use), two indicators of social conditions (number of groups encountered while hiking and the ability to camp out of sight and sound of other groups), and two indicators of managerial conditions (chance of receiv-

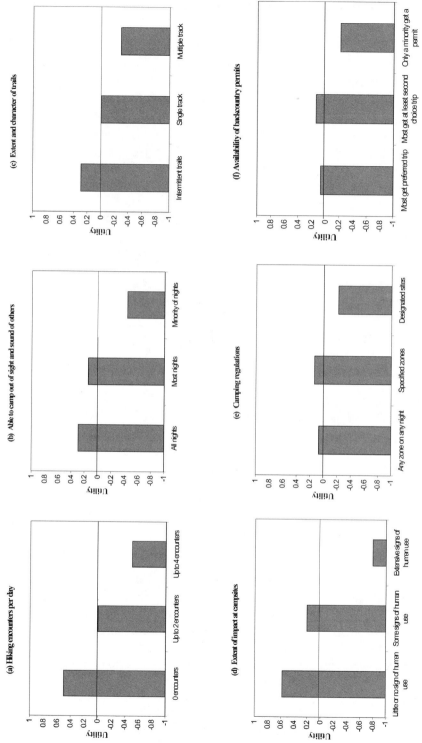

Figure 7.4. Denali wilderness setting indicators and standards and corresponding utility.

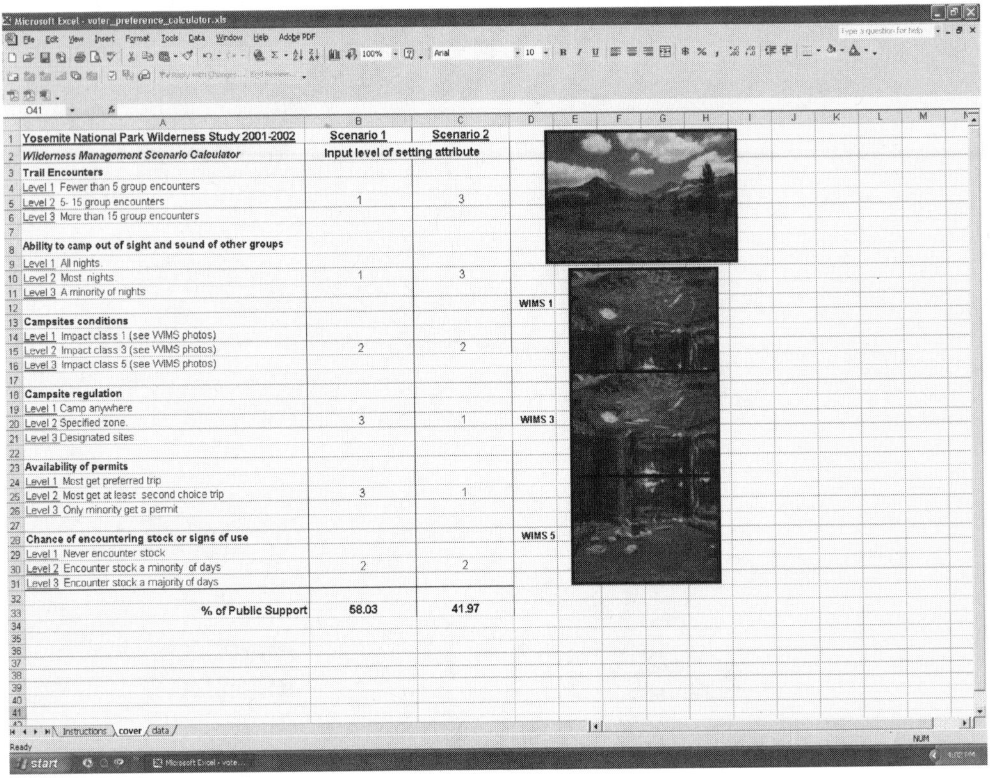

Figure 7.5. Wilderness Management Scenario Calculator.

ing a wilderness-use permit and degree of regulation of campsite location). Three standards (low, medium [or current], and high conditions) were used for each indicator variable.

Utilizing the coefficients derived from the study, visitor acceptability for any management scenario involving the six indicators and associated standards can be tested. To illustrate, consider the two wilderness management scenarios shown in figure 7.5. Scenario 1 can be referred to as the "solitude scenario," and Scenario 2 as the "freedom scenario." The numbers in each cell represent the level (low level=1, medium/current level=2, and high level = 3) of each indicator chosen to define the scenarios. The "solitude scenario" specifies that wilderness visitors would encounter fewer than five groups of people per day on the trail and have the ability to camp out of sight and sound of other groups all nights. However, visitors are required to camp in designated campsites, and only a minority receive a wilderness permit. The "freedom scenario" specifies that visitors would encounter more than fifteen groups per day along trails, and have the ability to camp out of sight and sound of other groups only a minority of nights. However, visitors can camp wherever they choose, and most visitors receive a permit for their preferred trip. In both of these scenarios, campsite impacts and the chance of encountering stock and associated impacts were held at the medium/current levels. The model estimates that nearly 60% of Yosemite wilderness visitors would prefer the "solitude scenario." This suggests that the majority of Yosemite wilderness visitors prefer to trade off some freedoms in order to attain a higher degree of solitude.

The wilderness management scenario calculator empowers managers to test all permutations of study variables and explore the relationships among the social, ecological, and managerial attributes (or indicators) that contribute to park and wilderness experiences. Managers who better understand these relationships can make more informed decisions concerning the development of indicators and associated standards, and carrying capacity more generally.

CHAPTER 8

Computer Simulation
Modeling of Visitor Use

It is when the harder decisions are made explicit that the arguments begin.

Simulation modeling is the imitation of the operation of a real-world process or system over time (Banks and Carson 1984; Law and Kelton 1991; Pidd 1992; Wang and Manning 1999). Simulation modeling enables the study of, and experimentation with, the internal workings of a complex system (e.g., a dispersed recreation setting, such as a national park). This approach is especially suited to tasks that are too complex for direct observation, manipulation, or even mathematical analysis.

The first generation of computer simulation modeling applications to parks and outdoor recreation, which came to be known as the *Wilderness Travel Simulation Model (WTSM)*, was introduced in the 1970s and continued through the mid-1980s (Borkan and Underhill 1989; Manning and Potter 1984; McCool et al. 1977; Potter and Manning 1984; Schechter and Lucas 1978; Smith and Headly 1975; Smith and Krutilla 1976; Underhill et al. 1986; van Wagtendonk and Cole 2005). The WTSM was designed to provide estimates of the number of encounters between recreation groups in a park or wilderness area, including their type (e.g., meeting, overtaking, encounters among different types of user groups), and location. Despite the early success of the WTSM, it fell into disuse largely due to the cost and technical difficulty of running computer simulations (Cole 2002).

Recent advances in computing technology have made computer simulation modeling more accessible and affordable (Pidd 1992). With improved computer simulation capabilities, a second generation of applications of computer simulation modeling to park and outdoor recreation management has emerged (Wang and Manning 1999; Cole 2005). This new generation of simulation modeling has been applied in several national parks and related areas to track visitor-use patterns and to assist managers in monitoring and managing carrying capacity and related issues (Daniel and Gimblett 2000; Gimblett et al. 2000; Manning et al. 1998a; Wang and Manning 1999; Wang et al. 2001; Cole 2005).

Simulation modeling has many potential applications in park and outdoor recreation planning and management. For example, simulation models of visitor use can provide detailed estimates of the amount and type of visitor use in a park, modeling its spatial and temporal distribution. In parks where visitor use is often dispersed over relatively large areas, and where visitor use can be difficult to observe directly, this type of information can be helpful in planning and managing such use (Cole et al. 2005). However, in the context of carrying capacity, simulation modeling can be especially helpful in three ways: monitoring indicator variables, estimating maximum visitor-use levels without violating crowding-related standards, and testing the effectiveness of management actions designed to maintain standards.

Monitoring Indicator Variables

Chapters 2 and 3 described the contemporary frameworks, such as Limits of Acceptable Change (LAC) and Visitor Experience and Resource Protection (VERP), that have been developed to guide analysis and management of carrying capacity, along with indicators and standards that are vital components of these frameworks. Once indicators and standards have been formulated, indicators must be monitored and management actions taken when monitoring data suggest that standards are in danger of being violated.

Monitoring indicator variables can be time consuming and costly. Moreover, some indicators, such as trail and campsite encounters, can be inherently difficult to observe. For these reasons, simulation models offer a potentially attractive alternative to on-the-ground monitoring. Once a simulation model is developed, it can be used to estimate the condition of indicator variables.

For example, a simulation model of visitor use of the carriage roads in Acadia National Park, Maine, was developed to help monitor the indicator variable of *persons-per-viewscape (PPV)* (the number of people at any one time on a typical hundred-meter section of the carriage road system) (Manning et al. 1998a; Wang and Manning 1999; Jacobi and Manning 1999; Manning and Wang 2005). The model was constructed using diary reports by visitors of their travel routes and times along the carriage roads, and counts of the number of visitors entering each of the eleven major access points into the carriage road system. These and related data were processed using the commercially available, general purpose, simulation software, *Extend*. The model was designed to estimate PPV levels along the carriage roads and can be run at any total daily-use level of the carriage road system. The park's monitoring program measures total daily use of the carriage roads through an electronic trail counter and uses the simulation model to estimate PPV levels (the crowding-related indicator variable) to ensure that crowding-related standards are not violated.

"Proactive" Monitoring

Contemporary carrying capacity frameworks such as LAC and VERP might be described as "reactive" in nature, at least in terms of monitoring and the management implications of resulting data. That is, management actions are taken only when monitoring data suggest that standards for indicator variables have been violated or are in danger of being violated. Carrying capacity frameworks could be applied more "proactively" by estimating the level of visitor use that will ultimately cause standards to be violated. Simulation modeling of visitor use can be used to make such estimates.

For example, a simulation model of visitor use at Arches National Park, Utah, was developed as part of a research program to help support application of the VERP framework (Lawson et al. 2003b; Wang et al. 2001). Initial phases of this research program were used to help formulate a suite of crowding-related indicators and standards throughout the park. For example, at Delicate Arch, an icon feature of the park, a crowding-related standard of thirty people-at-one-time (PAOT) was set (as described in chapter 6). To account for occasional random surges in visitation that are unavoidable, the standard was stated so that PAOT at Delicate Arch should not exceed thirty more than 10% of the time (National Park Service 1995).

A variety of methods were used to gather data needed to develop the simulation model of visitor use at Arches. A traffic counter placed at the entrance to the park was used to record the number of vehicles entering the park and the time each vehicle entered. These traffic data were collected during a seven-day period.

Data concerning visitor characteristics and their travel patterns within Arches National Park were collected through a series of on-site surveys administered to park visitors during the summer. Vehicle travel route questionnaires were administered to visitor groups in private automobiles and to tour bus drivers as they were exiting the park. Each respondent was asked to report their group's size, the amount of time they had spent traveling on the park roads, and where and how long they paused (for a minimum of five minutes) during the visit. Finally, with the aid of the interviewer, they were asked to retrace the route of their trip on a map of the park. Safety concerns preempted stopping cars and tour buses for surveying after dark, therefore, each sampling day ended at dusk.

A second questionnaire was administered to visitor groups returning from their hikes to Delicate Arch. One visitor from each group was asked to report the group's size, the amount of time they had spent on the trail to Delicate Arch and at the arch, and where and how long they paused (for a minimum of five minutes) during the hike.

Data needed to validate the output of the simulation model were gathered through a series of vehicle counts conducted at selected parking lots in the park. The number of vehicles in the Delicate Arch, The Windows, and Devil's Garden parking lots (the park's three major attraction areas) were counted eleven times a day on four days. The total number of vehicles entering the park was recorded with traffic counters on each of the days that parking lot counts were conducted.

The simulation model was built using the software Extend. The structure of the model consists of hierarchical blocks that represent specific parts of the park's road and trail systems, including entrance, intersection, road and trail, parking lot, and attraction site blocks, as shown in figure 8.1.

The primary purposes of the hierarchical entrance block are to generate simulated visitor groups and assign values for a set of characteristics or attributes that each visitor group "carries with them" through their simulated park visit (figure 8.1[a]). The rate at which visitor groups are generated by the model is determined by traffic counter data contained in the "vehicle generator" and "headways" blocks. The blocks labeled "attributes" and "travel routes" assign attribute values to newly generated visitor groups. The values of each group's attributes direct their travel through the simulated park visit and include travel mode (automobile or bus), group size, travel speed, and travel route. The range and frequency of attribute values used in the model are based on data collected in the visitor surveys. For example, travel speeds assigned to simulated visitor groups are based on the distance of travel routes reported in the visitor surveys and the total amount of time visitors spent traveling their routes. After receiving attribute values, the simulated visitor groups are directed to the "entrance queue" block where they are held for a period of time designed to simulate the waiting line that develops at the park entrance gate during the course of the day. The distribution of waiting times assigned to simulated visitor groups is based on actual waiting time data collected at the park's entrance gate during the peak period of the season. From the "entrance queue" block, simulated visitor groups enter the park's road network.

Road and trail section blocks simulate travel along park roads and trails (figure 8.1[b]). Simulated visitor groups enter a road or trail section through the "groups in" block. The amount of time groups spend along a road or trail section is determined by their travel speed and the length of the road or trail section. Like road and trail section blocks, parking lot blocks hold simulated visitor groups for periods of time based on data collected from the visitor surveys. Parking lot blocks also output the number of visitor groups parked in the lot throughout the simulated day.

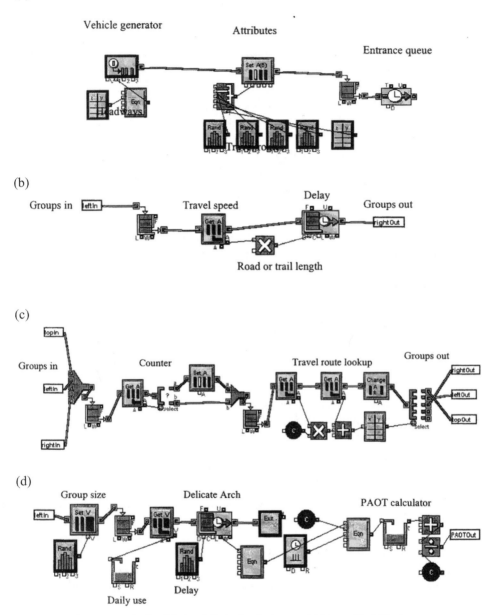

(a)

Vehicle generator

Attributes

Entrance queue

Headways

Tract or...

(b)

Groups in

Travel speed

Delay

Groups out

Road or trail length

(c)

Groups in

Counter

Travel route lookup

Groups out

(d)

Group size

Delicate Arch

PAOT calculator

Daily use

Delay

Figure 8.1. Arches National Park simulation model hierarchical blocks:
(a) entrance blocks, (b) road and trail section blocks, (c) intersection blocks,
(d) attraction site blocks—Delicate Arch.

Intersection blocks direct simulated visitor groups in the direction of their assigned travel routes when they arrive at road or trail intersections (figure 8.1[c]). The intersection blocks are used within the simulation model for road intersections, road and parking lot intersections, and trail intersections. Simulated visitor groups enter an intersection at the "groups in" block. The block labeled "counter" keeps track of the number of previous times, if any, the group has passed through the intersection. A series of blocks, unique for each intersection and labeled "travel route lookup," direct visitor groups to the next park feature based on the group's travel route and the number of previous times, if any, the groups have passed through the intersection. Visitor groups are sent to the next location through the "groups out" block.

Attraction site blocks disaggregate vehicle-based visitor groups into sets of individual hikers and output PAOT at the attraction site throughout the simulated day (figure 8.1[d]). Visitor groups "leave their vehicles" at the "group size" block, where they are assigned a group size, thus converting them into sets of individual hikers. The individual hikers are then directed to the attraction site, labeled "Delicate Arch" in this example, and delayed at the site for a period of time derived from the visitor survey data. The blocks labeled "PAOT calculator" calculate the percentage of time the number of people at Delicate Arch (i.e., the number of simulated hikers within the "Delicate Arch" block) exceeds thirty during the simulation period.

To estimate a daily carrying capacity for Delicate Arch, a series of simulations was run in which the total number of visitors hiking to the arch was varied. The average percentage of time that PAOT at Delicate Arch exceeded thirty (i.e., the maximum acceptable level of PAOT at Delicate Arch as determined from the normative and visual research methods described in chapters 5 and 6) was recorded for each use level modeled. An iterative process of increasing or decreasing the daily number of visitors hiking to the arch was followed until PAOT at Delicate Arch exceeded thirty an average of 10% of the time. For example, a series of twelve simulations was run for selected level of visitor use and the average percentage of time that PAOT exceeded thirty was calculated from the simulation results. If PAOT exceeded thirty an average of more than 10% of the time, the next set of twelve simulations was run at a lower use level, while if PAOT exceeded thirty an average of less than 10% of the time, the next set of twelve simulations was run at a higher level of visitor use. This iterative process was repeated to estimate a daily carrying capacity for Arches National Park, except that rather than varying the number of people hiking to Delicate Arch, the total number of vehicles entering the park was varied.

The model estimates that a maximum of 315 people can be allowed to hike to Delicate Arch between the hours of 5:00 a.m. and 4:00 p.m. without exceeding thirty PAOT at Delicate Arch more than 10% of the time. Further, the model results suggest that a maximum of 750 vehicles can be allowed to enter

TABLE 8.1. Parking lot validation statistics for the simulation model of visitor use at Arches National Park, UT

Parking lot counts	T statistic
Windows	−3.00[a]
Delicate Arch	1.46
Devil's Garden	−0.28
Parkwide	−0.40

[a] statistically significant at $\alpha = 0.05$

the park between the hours of 5:00 a.m. and 4:00 p.m. without violating the standard of quality for PAOT at Delicate Arch.

To test the validity of the simulation model, another series of model runs was conducted. The number of vehicles entering the park in the model runs was varied to match the number of vehicles entering the park on the four days that parking lot counts were conducted. The model runs were repeated twelve times for each of the four use levels to capture stochastic variation. For each of the four total-use levels modeled, the average number of vehicles in selected parking lots was calculated and compared to the actual parking lot counts.

Table 8.1 presents validation results based on comparisons between actual parking lot counts and model estimates. The four days of actual parking lot counts were combined and a set of four t-tests were performed to test for statistically significant differences among observed data and model estimates at each of the three parking lots and parkwide. There were no statistically significant differences among observed data and model outputs, except at The Windows parking lot. These findings suggest that the model is providing valid estimates of use levels in the park with the possible exception of one area.

Testing the Effectiveness of Management Actions

As noted earlier, management actions are needed to maintain standards that have been set. But how effective are alternative management actions? Management actions are conventionally tested through on-the-ground experimentation, which may be costly and politically risky. Simulation modeling can be used to test the potential effectiveness of selected management actions within a more controlled, "laboratory" context.

For example, the simulation model of visitor use at Arches National Park described earlier was used to test the potential effect of implementing a manda-

TABLE 8.2. Estimates of daily carrying capacity of
Delicate Arch with mandatory shuttle system

Arrival interval (minutes)	Passengers	Estimated daily carrying capacity	% increase in carrying capacity
60	37	407	29
30	21	462	47
15	12	528	68

tory shuttle bus system for hikers to Delicate Arch (Lawson and Manning 2002a). Public transit systems can be used to control the number and timing of visits to park attraction sites to help ensure that crowding-related standards are maintained.

To test the potential effectiveness of a shuttle bus system, the simulation model was modified to deliver visitors to the Delicate Arch trailhead at regularly scheduled time intervals. Separate model runs were conducted to simulate alternative shuttle bus schedules designed to arrive at the Delicate Arch trailhead every fifteen, thirty, and sixty minutes. For each shuttle bus system simulated, the number of visitors riding the shuttle bus and hiking to the arch was varied to estimate the maximum number of visitors that could be allowed to hike to Delicate Arch without exceeding the crowding-related standard (no more than thirty PAOT more than 10% of the time) for the arch.

Results of simulation runs conducted to test the effect of implementing a mandatory shuttle bus system to Delicate Arch are reported in table 8.2. The data in the third and fourth columns suggest that the daily carrying capacity of the arch could be increased by 29% to 68% if visitors were required to ride shuttle buses to Delicate Arch. For example, the model estimates that a shuttle bus system designed to deliver visitors to Delicate Arch every sixty minutes would increase the daily carrying capacity of the Arch from 315 hikers to 407 hikers between the hours of 5:00 a.m. and 4:00 p.m. Further, the results suggest that smaller, more frequent shuttle buses would increase the daily carrying capacity of Delicate Arch to an even greater extent. These increases in carrying capacity are due to a more even distribution of visitors over the day.

PART III

Case Studies of Measuring and Managing Carrying Capacity

The research methods described in part 2 can be used to inform analysis and management of carrying capacity in parks and related areas. These theoretical and methodological approaches can help guide formulation of indicators and standards for desired future conditions of park resources and experiences, monitor indicator variables, and evaluate the effectiveness of management actions designed to ensure that standards are maintained.

These research approaches are being applied in many national parks and related areas to help support analysis and management of carrying capacity. Part 3 presents eight case studies that illustrate this work. These case studies address a diverse array of national park areas that range along many dimensions, including geographic location; type of area (e.g., resource-based park, recreation area, cultural area); backcountry/wilderness and frontcountry; high use and low use; and many types of facilities and uses, including iconic attraction sites, hiking and biking trails, campsites, scenic roads, and historic buildings. Parks addressed include Acadia National Park, Yosemite National Park, Isle Royale National Park, Boston Harbor Islands National Recreation Area, Alcatraz Island (Golden Gate National Recreation Area), Muir Woods National Monument, Zion National Park, and Mesa Verde National Park.

The diverse contexts represented in these case studies have resulted in identification of many potential indicator variables, including a number of manifestations of crowding, conflicting uses and associated behavior, recreation-related

99

impacts to trails and campsites, litter, vandalism, group size, availability of parking, and visitor-caused noise. Study findings also offer an empirical basis to help formulate standards for these and related indicators. These case studies offer examples of the ways in which research can support application of contemporary carrying capacity frameworks in national parks and related areas.

CHAPTER 9

Managing Recreation at Acadia National Park

The National Parks present another instance of the
working out of the tragedy of the commons.

Acadia National Park, Maine, is one of the most intensively used national parks in the United States. While recreational use (2.2 million visits annually) does not rise to the level of some of the "crown jewel" western national parks (Yellowstone National Park, for example, accommodates 2.9 million visits annually), visits to Acadia are concentrated on its comparatively small size of less than 50,000 acres. Yellowstone, by comparison, is spread across 2.2 million acres. Given the intensive character of visitor use at Acadia, it is vital to monitor and manage recreation use and its associated impacts to help ensure protection of important park resources and the quality of the visitor experience.

Acadia has undertaken an expanding program of recreation-related monitoring, management, and associated activities (Manning et al. 2006). This program has been guided by the Visitor Experience and Resource Protection (VERP) framework described in chapter 2. Application of this framework has been supported by a program of natural and social science research in the park. This program of research was initiated on the park's carriage road system and has since expanded to include most areas of the park. Research has provided an empirical basis for formulating indicators and standards, helped devise monitoring approaches, and tested the potential effectiveness of alternative management practices.

The Carriage Roads

The carriage roads are a system of more than fifty miles of unpaved roads constructed at the direction of John D. Rockefeller Jr. in the early 1900s and represent one of the park's most significant cultural and recreational resources. Originally built for horse-drawn carriages, the carriage roads are now used primarily for hiking and biking and have become extremely popular. However, increased use has created concern for the quality of the recreation experience on the part of both visitors and managers. In response to this concern, a program of research was initiated to help formulate indicators and standards for the carriage road experience (Jacobi and Manning 1999; Wang and Manning 1999; Manning et al. 1998a).

A first phase of research focused on identifying potential indicators. A survey of a representative sample of carriage road visitors was conducted. Using both open- and close-ended questions, as described in chapter 4, visitors were asked to indicate what added to and detracted from the quality of their experience on the carriage roads. Two types of indicators were identified. One was crowding-related and concerned the number of visitors on the carriage roads. The other was conflict-related and addressed several "problem behaviors" experienced on the carriage roads, including bicycles passing from behind without warning, excessive bicycle speed, people obstructing the carriage roads by walking abreast or stopping in groups, and dogs off leash.

The first phase of research also documented existing patterns of use on the carriage roads by asking visitors to report their route of travel along the carriage roads. Visitor attitudes toward a variety of potential management practices were also measured. The carriage roads currently support a diversity of recreation opportunities defined both spatially and temporally. Some areas and times are relatively heavily used while other areas and times accommodate relatively light levels of use. Despite the problem behaviors, most visitors supported maintaining the current mix of carriage road users—hikers, bikers, and (a small number of) equestrians. Based on these findings, park management decided to maintain a diversity of carriage road experiences by establishing and defining two types of recreation opportunity "zones" for the carriage roads by location and time. However, both of these zones would continue to accommodate all types of visitors. The two carriage road zones would be defined by the same indicators, but different standards would be set.

A second phase of research focused on helping to formulate standards for the indicators. This research also used a survey of a representative sample of carriage road visitors as well as a survey of residents of surrounding communities. The surveys adopted normative theory and related empirical techniques, and visual research methods, as described in chapters 5 and 6.

Because of the relatively large number of visitors on the carriage roads, crowding was measured in terms of persons-per-viewscape (PPV), incorporating a visually based measurement approach. The viewscape for the carriage roads (the length of carriage road that can be seen at any one time) averages approximately one hundred meters. A series of nineteen study photographs was prepared that showed a range of zero to thirty visitors on a typical hundred-meter section of the carriage roads. The photographs were prepared using digital photo-editing software as described in chapter 6, and sample photographs are shown in figure 9.1.

Visitors were shown the photographs in random order and asked to rate their acceptability on a scale from –4 ("very unacceptable") to +4 ("very acceptable"). Study findings are shown in figure 9.2. This social norm curve (as described in chapter 5) represents the aggregate acceptability ratings for the sample of visitors. The norm curve indicates that visitors generally find that it is acceptable to see up to fourteen PPV. However, the quality of the experience is very marginal in the upper portion of this range, and visitors prefer to see far lower PPV levels. Because the park's *General Management Plan* specifies that the carriage roads should provide an especially high-quality visitor experience, lower PPV levels were set as the crowding-related standards for the two opportunity zones on the carriage roads.

Standards were also set for the four problem behaviors described earlier. Visitors were asked to report the maximum number of times it would be acceptable to experience each of these behaviors during a trip on the carriage roads. The resulting norms were used as a basis of formulating standards. As with crowding-related standards, standards for problem behaviors were set somewhat lower than maximum acceptable norms to ensure a relatively high level of quality for the visitor experience. Different standards were set for the high- and low-use portions of the carriage roads to ensure that a diversity of experiences was maintained.

The VERP framework requires that indicators be monitored and that management actions be taken to maintain standards. This is an ongoing process on the carriage roads. PPV levels are monitored in several ways, including estimation by a simulation model of visitor use as described in chapter 8. This model was developed using the travel routes that visitors reported in the first visitor survey. The model can be run at any total daily-use level for the carriage roads and estimates the resulting PPV levels that will occur along the carriage roads. The model was validated by comparing model estimates to actual carriage road-use data. Total daily use of the carriage roads is measured by an electronic trail-use counter. To ground-truth PPV estimates derived from the simulation model, PPV levels are also periodically monitored through (1) a short visitor survey that asks respondents to indicate which of the study photographs looks most like the use levels they experienced, and (2) observations of PPV levels at

(a) 0 people   (b) 5 people

(c) 10 people   (d) 15 people

(e) 20 people   (f) 30 people

Figure 9.1. Acadia carriage road photographs.

selected sites by trained employees. Monitoring data suggest that PPV standards have not yet been violated.

Problem behaviors are monitored through a short visitor survey conducted every three years. Monitoring data suggest that some standards are in danger of being violated, and several management actions have been implemented to address this issue. These management actions include development of "rules of the road," which are posted at all carriage road entrances; a liaison with local biking and hiking groups; and "courtesy patrols" on the carriage roads to educate visitors about appropriate behavior. These are the types of management actions that respondents favored, as found in the first visitor survey.

Trails on Isle au Haut

Isle au Haut is a relatively remote island outpost of Acadia National Park. It is not part of the main Mount Desert Island portion of the park and is located approximately seven miles off the Maine coast. Most visitors access the area by scheduled boat service, and the island is increasingly popular for day hiking and camping. Increasing use prompted concern over the potential resource and experiential impacts of recreation, and a program of natural and social science research was conducted to support analysis and management of the carrying capacity of the island (Manning et al. 2006).

An element of the natural science research addressed recreation impacts on trails. The literature suggests that soil erosion and visitor-caused trails are common and potentially important recreation-related impacts (Hammitt and Cole 1998; Leung and Marion 2000). Trail conditions on Isle au Haut were assessed through inventory and monitoring methods (Leung and Marion 1999a; Leung and Marion 1999b; Leung and Mation 1999c; Marion and Leung 2001), and both trail erosion and visitor-caused trails were found to be problems on the island.

An associated program of social science research was also conducted on trail-related impacts on Isle au Haut. Trail erosion and visitor-created trails impact park resources, but these impacts can also degrade the quality of the visitor experience. At what point do such impacts become noticeable and unacceptable from an experiential perspective? To help answer this question, two series of computer-edited photographs were prepared illustrating a range of recreation-related trail impacts on Isle au Haut, including trail erosion and proliferation of visitor-created trails. The study photographs were constructed to represent the type and pattern of impacts found in the ecological assessment of trail conditions. The photographs for visitor-created trails are shown in figure 9.3. As described in chapters 5 and 6, these photographs were incorporated into a survey asking visitors to the island to rate their acceptability. The resulting norm curve for visitor-created trails is shown in figure 9.4. Study findings from both the natural and social science research are being used to formulate a set of indicators and standards for resource and social conditions associated with trail use on Isle au Haut.

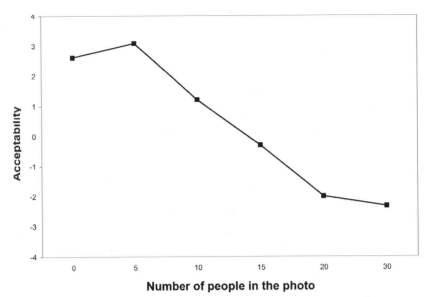

Number of people in the photo

Figure 9.2. Social Norm Curve for PPV on the Carriage Roads.

(a)   (b)

(c)   (d)

Figure 9.3. Visitor-created trail photos on Isle au Haut.

Scenic Roads

Automobile use is the primary means of transportation to and through most national parks, and for many visitors it is also the predominant means by which the parks are experienced. Some national park units, such as the Blue Ridge Parkway, are designed primarily for automobile use. Others, such as Acadia, rely on scenic roadways interspersed with pull-offs or spur roads to vistas to provide "the national park experience" to a majority of visitors. Despite the importance of automobile use in the national parks, little research has been conducted to examine crowding, automobile congestion, and carrying capacities on park roads. Therefore, research was needed to help formulate indicators and standards for automobile use on Acadia's scenic roads and to estimate the carrying capacity of the road system based on these standards.

The research described here was applied to the scenic drive through the Schoodic Peninsula section of Acadia (Hallo et al. 2005). The Schoodic Peninsula is one of three geographically separate areas of Acadia and is located approximately an hour's drive north from the main Mount Desert Island portion of the park. Frazer Point and Schoodic Point are the most frequently visited areas of the Schoodic Peninsula.

A scenic road is the sole travel route through the Schoodic Peninsula. The road enters the park near Frazer Point and follows the shoreline of the peninsula, reaching Schoodic Point prior to leaving the park. The road permits only one-way vehicle travel, except on access roads to both scenic points.

An initial visitor survey was conducted to identify potential indicators for the quality of the visitor experience at the Schoodic Peninsula. Both open- and close-ended questions probed for recreation-related impacts that detract from the quality of the visitor experience. The two most important indicators identified were crowding at park attractions, including the scenic road, and trail erosion.

A second visitor survey was conducted to gather data on standards for these indicator variables. (Since trail erosion at the Isle au Haut section of the park was addressed in the previous section, this section will focus on crowding on the scenic road.) Using the visual research approach described in chapter 6, a series of five computer-edited photographs of a section of the scenic road were prepared showing a range of zero to sixteen vehicles-per-viewscape (VPV). Study photographs are shown in figure 9.5. Using the normative research approach described in chapter 5, visitors were asked to judge the acceptability of each photograph on a scale from –4 ("very unacceptable") to +4 ("very acceptable"). Respondents were also asked to select the photograph that represented the other evaluative dimensions of "preference," "management action," and "displacement," as described in chapter 5. The same research methods were used to gather data on crowding-related standards for the attraction sites of Frazer Point and Schoodic Point. Study photographs of these sites showed a range of people-at-one-time

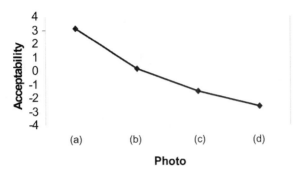

Figure 9.4. Social Norm Curve for Visitor-Created Trails
at Isle au Haut, Acadia National Park.

(PAOT). Findings from all study sites are summarized in table 9.1. For example, visitors reported that they would prefer to see an average of 2.5 VPV along the park's scenic road, and that they would no longer come back to the park if VPV reached an average of 12.7. Moreover, they felt the NPS should begin to limit visitor use of the scenic road when VPV reaches an average of 8.5.

A final element of research developed a simulation model of vehicle travel on the scenic road, as described in chapter 8. Primary input data were daily vehicle entry counts and travel routes collected from visitors. A representative sample of visitors was asked to participate in the study by carrying a Global Positioning System unit in their vehicle during their visit. The resulting simulation model was designed to estimate VPV levels on the scenic road based on the daily number of vehicles entering the park. The simulation model can be used to monitor VPV levels and estimate the maximum daily number of vehicles that can enter the park without violating crowding-related standards on the scenic road or at Frazer and Schoodic Points. The simulation model estimates that the number of vehicles entering the park per day could double without violating most of the

TABLE 9.1. Crowding-related standards at the
Schoodic Peninsula, Acadia National Park, ME

Normative standards	Scenic Road (VPV)	Frazer Point (PAOT)	Schoodic Point (PAOT)
Preference	2.5	35.3	22.6
Acceptability	7.5	85.0	70.1
Management action	8.5	89.0	71.2
Displacement	12.7	120.8	102.0

Note: VPV = vehicles/viewscape; PAOT = people-at-one-time.

(a) 0 vehicles

(b) 4 vehicles

(c) 8 vehicles

(d) 12 vehicles

(e) 16 vehicles

Figure 9.5. Study photographs of vehicle traffic on the Schoodic Peninsula scenic road.

crowding-related standards for this area. Moreover, crowding-related standards for the Schoodic Peninsula section of the park are likely to be violated first at Schoodic Point, so monitoring should be focused at this site.

CHAPTER 10

Day-Use Social Carrying
Capacity of Yosemite Valley

There is only one Yosemite Valley.

Yosemite Valley is the scenic heart of Yosemite National Park, California, arguably the first national park in the United States, and certainly one of America's best known and most popular national parks (Runte 1997; Runte 1990). Yosemite Valley is a glacially carved area of approximately seven square miles and features sheer granite walls of up to five thousand feet and several of the world's highest waterfalls. Yosemite National Park draws over four million visitors annually, and carrying capacity has been a long-standing and controversial issue. Day-use carrying capacity is of special concern because overnight accommodations in Yosemite Valley are fixed at current levels by park policy. Therefore, future increases in visitor use in Yosemite Valley will be driven by expanding day use.

A program of research was undertaken to help analyze and manage social carrying capacity in Yosemite Valley (Manning et al. 2003). This research was designed to support application of the Visitor Experience and Resource Protection (VERP) framework described in chapter 2. Three components of this work were conducted. The first helped provide an empirical foundation to identify and formulate indicators and standards for the visitor experience. The second component developed and applied a simulation model of visitor use at selected sites within Yosemite Valley to estimate the maximum number of daily visitors that could be accommodated without violating a potential range of crowding-related standards. Finally, a park exit survey was conducted to determine how many day-use visitors could be allowed to enter Yosemite Valley

110

each day without violating crowding-related standards at selected sites within the valley.

Indicators and Standards for the Visitor Experience

The first component of research was designed to help formulate indicators and standards for the quality of the visitor experience at strategic sites in and around Yosemite Valley. Seven sites were studied, including the base of Yosemite Falls, the trail to the base of Yosemite Falls, the base of Bridalveil Fall, the trail to the base of Bridalveil Fall, the trail to Vernal Fall, the trail to Mirror Lake, and Glacier Point. These sites were chosen in conjunction with park staff and represent a diversity of places that are important to visitors and to the park.

At each of these locations a survey of a representative sample of visitors was conducted. Questionnaires administered at these sites addressed both indicators and standards. Indicators were addressed through a series of open- and close-ended questions. Open-ended questions, as described in chapter 4, probed respondents for what added to or detracted from the quality of their visit to Yosemite Valley. Questions included: What have you enjoyed *most* about your visit to Yosemite Valley today? What have you enjoyed *least* about your visit to Yosemite Valley today? and If you could ask the National Park Service to improve some things about the way visitors experience Yosemite Valley, what would you ask managers to do? (Italics added.)

Close-ended questions asked respondents to rate the seriousness of several potential problem issues, as described in chapter 4. Using a response scale that ranged from 1 ("not a problem") to 3 ("big problem"), respondents were asked to rate the seriousness of the following potential problems: traffic congestion on roads, difficulty finding a parking place, inconsiderate drivers, too many tour buses, too many people on trails, too many people at places like the base of Yosemite Falls, too much noise, and too many rules and regulations.

Crowding-related issues, particularly the number of visitors on trails and at attraction sites, emerged as potentially important indicators. In the open-ended questions, crowding was reported by respondents as the least enjoyable aspect of their visit to all study sites. Moreover, in the close-ended questions, between 45.8% and 68.1% of respondents across the seven study sites judged "too many people on the trails" to be a "small" or "big" problem, and between 49.4% and 59.2% of respondents judged "too many people at places like the base of Yosemite Falls" to be a "small" or "big" problem.

Research on standards focused on crowding-related issues, including the number of people on trails and at attraction sites. A series of questions measured visitor crowding norms, as described in chapter 5. Since use levels are relatively

(a) 0 people

(b) 10 people

(c) 20 people

(d) 30 people

(e) 40 people

(f) 50 people

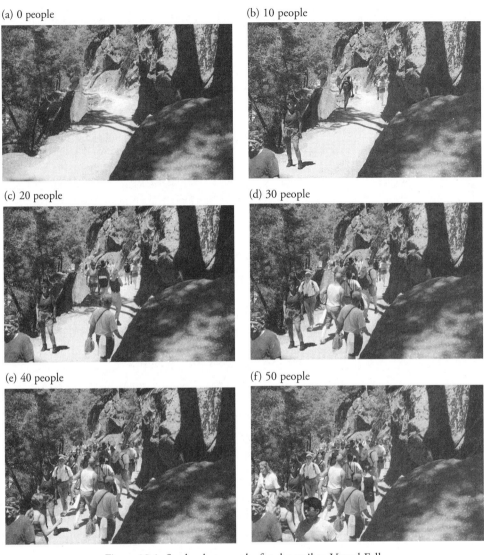

Figure 10.1. Study photographs for the trail to Vernal Fall.

high in Yosemite Valley, a visual approach, as described in chapter 6, was used to measure crowding norms. A series of computer-edited photographs was prepared for each study site showing a range of visitor-use levels. Study photographs are shown in figures 10.1 through 10.7. Respondents for each study site were asked to rate the acceptability of each photograph for that site on a scale that ranged from –4 ("very unacceptable") to +4 ("very acceptable"). In addition, respondents were asked to judge the photographs using several other dimensions of

(a) 0 people

(b) 16 people

(c) 32 people

(d) 48 people

(e) 64 people

(f) 80 people

Figure 10.2. Study photographs for the trail to Yosemite Falls.

evaluation to gain additional insights into how respondents judge alternative use levels, as described in chapter 5. These included "preference," "displacement," and "management action."

Data on crowding-related standards are shown in table 10.1. "Acceptability"-based standards were derived by plotting average acceptability ratings for each of the visitor use levels shown in the six photographs for each study site. People-at-one-time (PAOT) and persons-per-viewscape (PPV) standards

(a) 0 people

(b) 36 people

(c) 72 people

(d) 108 people

(e) 144 people

(f) 180 people

Figure 10.3. Study photographs for the base of Yosemite Falls.

(a) 0 people

(b) 6 people

(c) 12 people

(d) 18 people

(e) 24 people

(f) 30 people

Figure 10.4. Study photographs for the trail to Bridalveil Fall.

shown in the table are points at which average acceptability ratings cross the zero or neutral point on the acceptability scale (i.e., fall out of "acceptable" range and into the "unacceptable" range). "Preference"-, "management action", and "displacement"-based standards were derived by calculating the average number of people in the photographs selected by visitors in response to these questions.

(a) 0 people

(b) 6 people

(c) 12 people

(d) 18 people

(e) 24 people

(f) 30 people

Figure 10.5. Study photographs for the base of Bridalveil Fall.

These data suggest a potential range of standards for each study site. For example, visitors to Yosemite Falls prefer to see an average of 43 PAOT in the viewing area at the base of the falls, and would no longer visit this area if they saw an average of 126 PAOT. Moreover, they think the NPS should begin to limit use of this area when PAOT reaches an average of 100.

As discussed in chapter 5, none of the points within these ranges for each study site is more "valid" than any other. Each point has potential strengths and weaknesses. For example, standards based on preference-related norms may

(a) 0 people

(b) 14 people

(c) 28 people

(d) 42 people

(e) 56 people

(f) 70 people

Figure 10.6. Study photographs for Glacier Point.

(a) 0 people

(b) 9 people

(c) 18 people

(d) 27 people

(e) 36 people

(f) 45 people

Figure 10.7. Study photographs for the trail to Mirror Lake.

result in very high-quality recreation experiences but would restrict access to a relatively low number of visitors. In contrast, standards based on acceptability, management action, or displacement allow access to greater numbers of visitors but may result in recreation experiences of lesser quality. Findings that offer insights into multiple evaluative dimensions provide a potentially rich base of information and may lead to formulation of the most thoughtful and informed

TABLE 10.1. Alternative crowding-related standards for study sites at Yosemite National Park, CA

Normative standard	Trail to Vernal Fall (PPV)	Trail to Yosemite Falls (PPV)	Base of Yosemite Falls (PAOT)	Trail to Bridalveil Fall (PPV)	Base of Bridalveil Fall (PAOT)	Glacier Point (PAOT)	Trail to Mirror Lake (PPV)
Preference	11	18	43	7	8	19	10
Acceptability	26	40	92	18	20	42	24
Management action[a]	30	46	100	20	19	49	26
Displacement[b]	39	60	126	26	25	61	34

Note: PPV = persons/viewscape; PAOT = people-at-one-time.

[a] Number of respondents reporting that the NPS should not limit use

Trail to Vernal Fall = 66   Base of Bridalveil Fall = 88
Trail to base of Yosemite Falls = 59   Glacier Point = 97
Base of Yosemite Falls = 57   Trail to Mirror Lake = 65
Trail to Bridalveil Fall = 73

[b] Number of respondents reporting that none of the photographs were so unacceptable that they would no longer visit this site

Trail to Vernal Fall = 37   Base of Bridalveil Fall = 113
Trail to base of Yosemite Falls = 40   Glacier Point = 99
Base of Yosemite Falls = 46   Trail to Mirror Lake = 65
Trail to Bridalveil Fall = 95

standards. Moreover, such data allow more explicit understanding of the potential tradeoffs between use level and quality of the recreation experience.

Simulation Models

The second component of research focused on developing a series of computer-based simulation models of visitor use at all study sites. The models were built using the commercial simulation software package Extend, as described in chapter 8. Model input was based on detailed counts and observations of visitor use at each study site. Variables included length of trails, length of typical trail viewscapes (how far along the trail a hiker can typically see), number of visitors arriving per hour, visitor group size, length of time visitors stop at attractions, and the speed at which visitors hike trails. Model output was PAOT at attraction sites and PPV along trails, corresponding to the crowding-related standards measured using the study photographs described earlier. Resulting simulation models were designed to estimate PAOT and PPV for alternative total daily-use levels at each study site. In this way, maximum total daily-use lev-

TABLE 10.2. Range of daily carrying capacities
of study sites at Yosemite National Park, CA

Normative standard	Trail to Vernal Fall (PPV)	Trail to Yosemite Falls (PPV)	Base of Yosemite Falls (PAOT)	Trail to Bridalveil Fall (PPV)	Base of Bridalveil Fall (PAOT)	Glacier Point (PAOT)	Trail to Mirror Lake (PPV)
Preference	2,100	4,500	3,000	1,200	700	1,500	1,800
Acceptability	6,000	9,500	5,500	3,200	1,700	4,000	5,000
Management action	7,000	11,000	5,900	3,500	1,700	4,800	5,500
Displacement	9,300	13,000	7,300	4,800	2,300	6,300	7,700

Note: PPV = persons/viewscape; PAOT = people-at-one-time.

Daily carrying capacities were calculated to allow normative standards to be exceeded a maximum of 10% of the time.

els without violating alternative standards could be estimated for each study site. In effect, these maximum total daily-use levels are estimates of social carrying capacity for each study site.

All models were run multiple times to average out the randomness associated with each individual model run. The range of daily carrying capacities for each study site is shown in table 10.2. (These estimates were calculated to allow crowding-related, normative standards to be violated a maximum of 10% of the time, an issue discussed in chapter 5.) Daily carrying capacities can be seen to vary substantially across the range of normative standards and across study sites. For example, the simulation model for the base of Yosemite Falls estimates that this area could accommodate 3,000 visitors a day using the preference-based standard of 43 PAOT, but could accommodate 7,300 visitors a day using the displacement-based standard of 126 PAOT.

Percentage of Day Users

The final component of research focused on applying carrying capacity estimates specifically to day users. As noted earlier, overnight visitor capacity of Yosemite Valley is fixed at current levels by park policy. Application of carrying capacity estimates to day users required two types of information. First, the percentage of day users at each study site was determined by a short series of questions in the visitor questionnaires described earlier. Respondents were simply asked if they had spent last night or were planning to spend tonight in Yosemite Valley. Second, the percentage of day visitors to Yosemite Valley who visit each of the study sites was determined through a park exit survey. Visitors exiting the park were

TABLE 10.3. Percentage of day users at
study sites at Yosemite National Park, CA

	Trail to Vernal Fall (PPV)	Trail to Yosemite Falls (PPV)	Base of Yosemite Falls (PAOT)	Trail to Bridalveil Fall (PPV)	Base of Bridalveil Fall (PAOT)	Glacier Point (PAOT)	Trail to Mirror Lake (PPV)
Percentage of visitors at each study site who were day users	42.9	36.8	36.8	77.0	77.0	62.8	37.4
Percentage of Yosemite Valley day users who visited each study site	10.2	35.4	35.4	43.4	43.4	18.1	8.4

Note: PPV = persons/viewscape; PAOT = people-at-one-time.

randomly selected and administered a short questionnaire that asked if they were day visitors to the park and, if so, had they visited each of the study sites. Findings are shown in table 10.3. It is clear these figures vary substantially from site to site. For example, 77% of visitors to the base of Bridalveil Fall are day visitors compared to 36.8% at the base of Yosemite Falls. Moreover, only 8.4% of all day visitors to Yosemite Valley hike the trail to Mirror Lake compared to 35.4% that visit the base of Yosemite Falls.

Day-Use Carrying Capacity

Based on the three components of research, a range of daily day-use social carrying capacities can be estimated for Yosemite Valley. These estimates are the maximum daily number of day-use visitors that can be accommodated in Yosemite Valley without violating PPV or PAOT standards at each study site. The estimates are based on findings from the three components of research and are applied in three steps or mathematical calculations: First (step 1), the number of overnight visitors in Yosemite Valley at all study sites was estimated using findings from the visitor surveys and the counts of visitor use. Second (step 2), the number of overnight visitors to each study site was then subtracted from the maximum total daily use levels that can be accommodated at each study site. This leaves the maximum number of day visitors that can be accommodated at each study site. Third (step 3), this number was then expanded to account for the fact the park exit survey found that only certain percentages of all day visitors to Yosemite Valley visit each of the study sites.

An example will help illustrate how these estimates were calculated. Table 10.1 indicates that the "preference-based" crowding standard along the trail to Bridalveil Fall is 7 PPV. This number was derived from the survey of visitors to this site and respondents' assessment of the study photographs showing a range of visitor-use levels. Table 10.2 indicates that a maximum of 1,200 visitors could use this trail per day without violating this preference-based standard. This number was derived from the computer-based simulation model of this site. The three steps or mathematical calculations are then applied to derive the estimate of the daily day-use carrying capacity of Yosemite Valley based on the findings for the trail to Bridalveil Fall. The number of overnight visitors at Bridalveil Fall was estimated (Step 1). This was done by multiplying the percentage of day users at this site (77%) (taken from the first row of table 10.3) by the average daily number of all visitors to this site (3,501) (taken from the counts used to develop the computer simulation model of visitor use). From this operation, it is estimated that an average of 805 visitors to the trail to Bridalveil Fall are overnight visitors to Yosemite Valley on an average summer day. This leaves a capacity of 395 day visitors at this site (1,200 minus 805) (Step 2). Finally, the number of day users that can be accommodated on the trail to Bridalveil Fall needs to be expanded to estimate the number of day users that can be accommodated in Yosemite Valley (Step 3). Since only 43.4% of all day users who visit Yosemite Valley visit the trail to Bridalveil Fall (taken from the second row of table 10.3), this means that 910 day users can be accommodated in Yosemite Valley (395 divided by .434) without violating the preference-based standard for the trail to Bridalveil Fall.

Issues in Applying Carrying Capacity

The program of research and its application described in this chapter raises a number of issues regarding estimation and management of carrying capacity of parks and related areas. As discussed in part I of this book, an emerging principle of carrying capacity is that it must be guided by management objectives/desired conditions and associated indicators and standards. A corollary of this principle is that there is no one, inherent, carrying capacity of a park or recreation area. Rather each park (or even site within a park) has a range of capacities depending upon the degree of resource protection and type of recreation experience to be provided.

Data developed in this study illustrate this point. The number of people encountered at all study sites is important to most visitors, therefore measures of PPV and PAOT are good indicators of the quality of the visitor experience.

However, standards for this indicator vary substantially depending upon the evaluative dimension used in the study (i.e., the type of recreation experience). Crowding norms ranged consistently (with one small exception) from a low associated with "preference" (a high-quality experience) to a high associated with "displacement" (a lower-quality experience) for all study sites. The carrying capacities of these sites, and ultimately Yosemite Valley, vary accordingly.

A related principle of carrying capacity is that some element of management judgment must be exercised. Again, the data developed in this study illustrate this principle. What point (or points) along the range of standards and associated carrying capacities should be selected for management purposes? This is ultimately a judgment that should consider a variety of other factors inherent in carrying capacity, including the purpose and significance of the area (as may be defined in law or policy), the fragility of natural and/or cultural resources, financial and/or personnel resources available for management, historic precedent, and interest-group politics. However, such judgments should be as empirically informed as possible. Moreover, management judgments about standards and associated carrying capacities are not necessarily "either/or" decisions. In fact, it may be highly desirable to provide a spectrum of recreation opportunities within parks and among parks within a region.

These considerations may suggest valid reasons to formulate certain standards that in turn define carrying capacities. For example, there may be valid reasons for the trail to Mirror Lake to be managed for a PPV standard that is close to the "preference" end of the range of crowding norms identified in this study. This trail is less accessible than the other trails included in this study, has relatively low historic-use levels, and traverses areas containing fragile cultural resources. The trail to the base of Yosemite Falls, however, is easily accessible, highly used, does not contain especially sensitive natural or cultural resources, and leads to an "icon" feature of the park that most visitors want to experience. Thus, it may be reasonable to manage this trail for a PPV standard that is more toward the "management action" or even "displacement" end of the range of standards, and therefore for a relatively high carrying capacity. In this way, management judgments can lead to a spectrum of recreation opportunities that serve a diversity of public desires and appropriately balance competing carrying capacity considerations. However, such management judgments should be as informed as possible, based on data such as those developed in this study, and fashioned deliberately within the structured, rational context of a framework such as VERP. This approach to carrying capacity is most likely to lead to thoughtful park and outdoor recreation management that serves the needs of society and can withstand the inevitable test of public scrutiny.

The carrying capacity-related data gathered in this study has potentially important implications for other components of park and outdoor recreation management. For example, Yosemite National Park is one of many national parks being considered for new or expanded public transit systems. Data on social carrying capacity of alternative sites within Yosemite Valley are instrumental in designing a transit system that will deliver the "right" number of visitors to the "right" locations at the "right" times.

CHAPTER 11

Wilderness Camping
at Isle Royale National Park

The values that visitors seek in the parks are steadily eroded.

Since passage of the federal Wilderness Act in 1964, recreational use of wilderness has grown steadily and continues to be on the rise, particularly in national parks (Cole 1996). The wilderness portion of Isle Royale National Park, Michigan, is a good example: approximately 99% of the park's land base is designated wilderness, and visitation to the park during the 1990s grew at a rate of 4% to 5% annually. On a per-acre basis, the park has the highest number of wilderness overnight stays in the national park system (Farrell and Marion 1998).

Growing demand for wilderness camping at Isle Royale has led to potential problems. During peak periods of the season, campground capacities are commonly exceeded, and some camping groups must share campsites. During the peak-use period in 1997, a survey of wilderness campers found that respondents reported having to share campsites nearly 20% of camping nights (Pierskalla et al. 1996; Pierskalla et al. 1997). Moreover, most campers surveyed indicated that having to double up with other camping groups detracted from the quality of their experience.

To support analysis and management of wilderness camping at Isle Royale, a two-phase program of research was developed and administered (Lawson and Manning 2003a; Lawson and Manning 2003b). This research program was designed to address both the *descriptive* and *evaluative/prescriptive* components of carrying capacity, as described in chapter 2. In the case of Isle Royale, one of the important descriptive issues to be addressed is the relationship between the number and distribution of wilderness camping groups and the percentage of groups that must share campsites. The prescriptive component concerns the seemingly more subjective issue of how much impact should be allowed. In the case of Isle Royale, for example, how much campsite sharing should be allowed before use limits or other management actions are taken? A computer-based simulation model (described in chapter 8) of wilderness camping at Isle Royale was developed in the first phase of research to address the descriptive component of carrying capacity and to help inform the second phase of research, which involved application of a stated choice-based visitor survey (described in chapter 7) to address the prescriptive component of carrying capacity.

Isle Royale National Park is located in the northwest corner of Lake Superior, approximate seventy-five miles from Houghton, Michigan, and twenty miles from Grand Portage, Minnesota. The park has a system of thirty-six campgrounds, with a total of 244 designated tent and shelter sites dispersed along a network of 165 miles of trails. Primary recreation activities at the park, which is open to visitors from mid-April until the end of October, include hiking and camping (Farrell and Marion 1998).

Visitors interested in wilderness camping at Isle Royale are required to obtain a permit. As part of the permitting process, visitors are asked to report their anticipated itinerary, identifying the number of nights they plan to be in the park and the campground they intend to stay at each night of their camping trip. However, visitors are not required to follow their proposed itinerary, and there are no restrictions on the number of permits issued for camping in the park. While visitors have the option to obtain special permits for off-trail hiking and camping, the vast majority choose to camp at the designated campground sites (Farrell and Marion 1998).

Isle Royale National Park's approach to wilderness camping management is designed to maximize public access to the park and to maintain visitors' sense of spontaneity and freedom. However, this management approach, coupled with increased wilderness visitation at the park, has resulted in campground capacities being exceeded during peak periods of the visitor-use season. Park managers decided to address this issue by formulating a standard for campsite sharing (i.e., the maximum amount of campsite sharing that will be allowed). As park staff attempt to identify an appropriate and feasible standard for campsite sharing, they are faced with a number of difficult questions. For example, to what extent would use limits or fixed itineraries need to be imposed in order to achieve alternative standards for campsite sharing? To what degree could provision of public access, protection of visitor freedoms, and reduced campground crowding be optimized by redistributing use temporally and/or spatially? Could alternative standards for campsite sharing be achieved by adding new campsites to the park rather than by limiting use? If so, how many additional campsites would be needed, and where would they need to be located? Embedded in all of these questions are tradeoffs among visitor freedoms, spontaneity of visitor experiences, public access, natural resource protection, and opportunities for camping solitude. This study uses simulation modeling and tradeoff analysis to assist managers in answering these and related questions.

Computer Simulation Model

A simulation model of wilderness camping at Isle Royale was developed using the commercially available, general purpose, simulation software Extend, as described in chapter 8 (Lawson et al. 2003a). Wilderness camping permits issued by the park provided the primary source of data needed to construct the simulation model. In particular, detailed information concerning the starting and ending date of each group's trip, camping itinerary, and group size were used as inputs to the simulation model. As mentioned earlier, visitors are asked to report their expected camping itinerary as part of the park's permitting process, but they are not required to follow their proposed itinerary. This policy raises concern that use of the permit data as an input into the simulation model may not be valid. That is, the travel routes reported on the permits may not accurately reflect the actual itineraries followed by wilderness camping groups. In order to address this concern, all wilderness camping groups during the period of this study were asked to correct the camping itinerary reported on their permit and to return the corrected permit to the visitor center at the end of their camping trip.

Data needed to validate the outputs of the simulation model were gathered through a series of campground occupancy observations conducted throughout the visitor-use season. Campground hosts at Belle Isle and Daisy Farm campgrounds counted the number of groups in each at the end of the day on randomly selected dates throughout the season. In addition to the counts conducted by campground hosts, wilderness rangers conducted campground occupancy observations throughout the park during patrols. Data collected by the wilderness rangers included the date, campground name, and number of groups in the observed campground.

An initial set of simulation runs was conducted to characterize the extent of campsite sharing during July and August and the remainder of the season (referred to as the July/August peak and the low-use period of the season, respectively). This was followed by a series of simulation runs conducted to estimate the effectiveness of alternative strategies for managing campsite sharing, including use limits, spatial and temporal redistribution of use, fixed itineraries, and campsite construction.

Finally, a set of simulations was run at current July/August peak wilderness-camping levels to validate simulation model output. Based on outputs from these simulations, averages were computed for the number of groups in selected campgrounds on weekend nights and on weekday nights. Simulation model output was compared to average campground occupancy data collected by park staff.

Model Output

As outlined earlier, the simulation model was designed to produce several types of output, including estimating the extent of current campsite sharing, testing the potential effectiveness of alternative management practices designed to reduce campsite sharing, and evaluating the validity of the model. These model outputs were used to identify a set of four feasible and realistic management alternatives, including their effect on campsite sharing and related issues, such as public access, visitor freedom, and environmental impacts associated with campsite development.

Extent of Campsite Sharing

Simulation model runs designed to estimate the extent of wilderness campsite sharing found that during the July/August peak, an average of nearly 9% of groups were required to share campsites per night. During the low-use period, the model estimates that, on average, less than 1% of groups were required to share campsites per night.

Evaluation of Management Practices

A series of simulation model runs was also conducted to test the potential effectiveness of several management practices designed to reduce campsite sharing. First, the model was used to estimate the extent to which camping permits would need to be reduced to ensure that no more than 5% of groups would have to share campsites per night (the 5% campsite-sharing standard was one of several standards park staff was considering). Results suggest that in order to achieve this standard for campsite sharing, the average number of permits issued per weekend day would have to be reduced by about one-quarter, and the number of permits issued per weekday would have to be reduced from about thirty-four to twenty-eight. Overall, the model estimates that the average number of permits issued per day would have to be reduced from about thirty-nine to about thirty-one. This would result in approximately 22% fewer wilderness camping permits available to visitors during the July/August peak.

Second, the model was used to estimate the extent to which spatial and temporal redistribution of wilderness trips might reduce campsite sharing. Nearly all wilderness trips start at one of two trailheads (Windigo and Rock Harbor); most start at Rock Harbor. Moreover, more trips start on weekend days than on weekdays. The model was run to test the effect on campsite sharing of redistributing use evenly between the two primary trailheads and weekend days and weekdays. However, these spatial and temporal redistributions of use did not substantially reduce campsite sharing (in fact, the spatial redistribution increased campsite sharing).

Third, output from the simulation model provided insight into the effect that requiring visitors to follow prescribed, fixed-camping itineraries during the July/August peak period would have on campsite sharing. The results of a series of simulation runs suggest that there are an average of 148 camping groups in the park per night during July and August. As mentioned earlier, there are a total of 244 wilderness campsites in the park. Theoretically, if the park instituted a fixed-itinerary permit system, wilderness camping use could increase by approximately 60% while virtually eliminating campsite sharing. This assumes, however, that wilderness camping groups would not deviate from their prescribed itineraries and that it would be possible to design a set of fixed itineraries that would result in every campsite being occupied.

Fourth, the model was used to test the effect of adding additional campsites on campsite sharing. The park's *General Management Plan* allows for construction of up to thirteen additional campsites within specified existing campgrounds. The simulation model estimates that, without instituting any limits on use, an average of just under 7% of groups would share campsites per night if the thirteen additional campsites were constructed.

Model Validation

No statistically significant differences were found between observed campground occupancies collected by park staff and simulation model output. While there may be a possibility of Type II error associated with these tests due to the relatively small sample sizes, there are no *substantive* differences between the observed campground occupancies and the corresponding model output. This suggests that the simulation model accurately represents wilderness camping conditions at the park and that the permit data are valid inputs to the model.

Management Alternatives

Development and application of the simulation model described earlier enabled managers and researchers to "test" the potential effectiveness of alternative management practices for reducing campsite sharing and enhancing camping solitude.

In addition to their effect on campsite sharing, each of these management practices also has effects on other desired attributes or indicators of the wilderness experience at Isle Royale. For example, assigning fixed camping itineraries would allow for a 30% increase in camping permits while reducing campsite sharing to less than 1% of groups per night at the current use level. However, this alternative would reduce visitor freedom of travel.

Another management alternative would involve building additional campsites. However, the simulation model estimates that an additional seventy campsites would have to be constructed to reduce campsite sharing to less than 1% of groups per night. Construction of seventy additional campsites would have potentially important environmental impacts on the wilderness. A third management alternative would be to reduce the number of camping permits to achieve a standard of no more than 5% of groups sharing a campsite per night. This alternative would not restrict visitor freedom of travel and would not entail the environmental costs of building additional campsites, but the number of camping permits would have to be reduced by 22%. A final management alternative would be to maintain the status quo—continue issuing an average of thirty-nine camping permits per day, do not build any new campsites, and do not require fixed camping itineraries—but this would mean that about 9% of groups per night would share campsites. These four management alternatives are summarized in table 11.1

Application of simulation modeling to wilderness camping at Isle Royale helps describe a set of feasible and realistic management alternatives, including

TABLE 11.1. Isle Royale wilderness management
alternatives quantified based on simulation model output

Wilderness indicator	Alternative 1 Status quo	Alternative 2 Permit quota	Alternative 3 Fixed itineraries	Alternative 4 Campsite construction
Public access	Current use (39 permits/day)	22% reduction in use (31 permits/day)	30% increase in use (52 permits/day)	Current use (39 permits/day)
Campsite construction	No new campsites	No new campsites	No new campsites	70 new campsites
Freedom of travel	No fixed itineraries	No fixed itineraries	Fixed itineraries	No fixed itineraries
Campsite sharing/ solitude	9% of groups share campsites/ night	5% of groups share campsites/ night	<1% of groups share campsites/ night[a]	<1% of groups share campsites/ night

[a] Assumes permits are issued to achieve 80% occupancy rate to adjust for noncompliance.

the associated tradeoffs among wilderness attributes or indicators. The management alternatives included in table 11.1 constitute an important part of the descriptive component of carrying capacity: they describe a reasonable range of possibilities. This descriptive information helps inform the more prescriptive component of carrying capacity: which of these alternatives is most desirable, and which attributes or indicators of the wilderness experience are most important?

Stated Choice Modeling

A visitor survey incorporating stated choice modeling (as described in chapter 7) was conducted at Isle Royale to assess visitors' support for the management alternatives presented in table 11.1. In the questionnaire, wilderness campers were presented with a series of paired comparison questions. Within each paired comparison, respondents were asked to indicate which of the two wilderness camping-management alternatives they preferred. Each management alternative included in the questionnaire was defined by four indicators describing (1) the extent of campsite sharing in the park, (2) the degree of regulation of camping itineraries, (3) the intensity of campsite development, and (4) the availability of wilderness camping permits. These indicators were selected to reflect the issues being considered in wilderness management at the park. The standards of the indicators were defined based on computer simulation model estimates of the extent to which campsite sharing could be reduced through various management actions. Table 11.2 lists the study indicators and standards used to define the wilderness camping management alternatives included in the paired comparison questionnaire.

The study indicators were combined at the standards outlined in table 11.2 into a set of sixteen paired comparisons based on a fractional factorial design (as described in chapter 7). To reduce respondent burden, the sixteen paired comparisons were blocked into two questionnaire versions, each containing eight pairwise comparisons (Green and Srinivasan 1978; Louviere et al. 2000). An example of a pairwise comparison question is presented in figure 11.1.

The paired comparison questionnaire was administered to randomly selected visitors at the completion of their wilderness camping trips. Study participants were randomly assigned to complete one of the two versions of the questionnaire on a laptop computer. For each question, respondents were asked to read through two management scenarios and indicate which they preferred. The response rate for the survey was 100%, resulting in a total of two hundred completed questionnaires (approximately one hundred respondents for each version of the questionnaire) and 1,600 pairwise comparisons.

Figures 11.2(a)–(d) plot the utility associated with each standard of the four

TABLE 11.2. Indicators and standards of camping-management scenarios used in the Isle Royale wilderness study

Percent of groups that share campsites per night
    0% of groups share campsites with other people not in their group/night
    5% of groups share campsites with other people not in their group/night
    10% of groups share campsites with other people not in their group/night

Construction of additional campsites
    No new campsites are constructed
    70 new campsites are constructed within existing campgrounds

Regulations concerning wilderness camping itineraries
    Visitors are required to follow prescribed, fixed itineraries
    Visitors are not required to follow fixed itineraries

Availability of wilderness camping permits
    The number of permits available to the public is increased by 30%
    The number of permits available to the public is held constant at current use levels
    The number of permits available to the public is reduced by 20%
    The number of permits available to the public is reduced by 75%

Management scenario *A*	Management scenario *B*
• *0% of parties* share campsites with other people not in their party per night.	• *5% of parties* share campsites with other people not in their party per night.
• *No new campsites* are constructed.	• *70 new campsites* are constructed within existing campgrounds.
• Visitors *are required* to follow prescribed, fixed itineraries.	• Visitors *are not required* to follow prescribed, fixed itineraries.
• The number of backcountry camping permits made available to the public during July and August is *held constant* at current use levels.	• The number of backcountry camping permits made available to the public during July and August is *increased by 30%*.

Please check one box

Prefer management scenario A ☐   Prefer management scenario B ☐

Figure 11.1. Example of pairwise comparison question.

study indicators (Nicholson 1995). Values on the *x*-axis represent the standard of the corresponding study indicator, and values on the *y*-axis represent utility. Standards for indicators with high utility values are preferred by visitors to standards for indicators with lower utility values. The plots provide graphic insight into the relative sensitivity of visitors to the standards of the management indicators. For example, utility drops sharply as limits on visitor use change from a permit quota that reduces visitor use by 20% (0.402) to one that reduces visitor use by 75% (–0.761), whereas the loss of utility is less dramatic (and not statistically significant) as visitor-use limits change from maintaining current use (0.577) to reducing use by 20% (0.402) (figure 11.2(d)). Similarly, if the park were to move from not requiring fixed itineraries to a fixed-itinerary system, utility would drop substantially, while the change in utility associated with an increase of campsite sharing from 0% of parties sharing campsites per night to either 5% or 10% of parties sharing campsites per night is moderate, and not statistically significant.

The parameters of the stated choice model were combined with descriptive data from the computer simulation model to predict visitors' support for the four wilderness management alternatives presented in table 11.1. In particular,

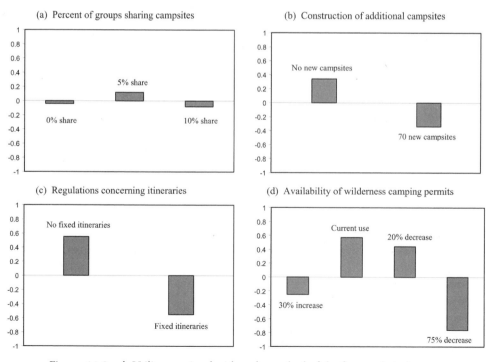

Figures 11.2a–d. Utility associated with each standard of the four study indicators.

indicator values corresponding to the four management alternatives were substituted into the stated choice model to predict the proportion of visitors that would support each alternative in a hypothetical referendum (Opaluch et al. 1993). The results of this analysis suggest that the greatest support among visitors is for the "status quo" and "permit quota" options, with 36% and 39% of visitors predicted to support each of these alternatives, respectively. While the "campsite construction" alternative is less popular than the "status quo" and "permit quota" alternatives, the stated choice model predicts that nearly 20% of visitors would support this option. The "fixed-itinerary" alternative is substantially less favorable to visitors than any of the other alternatives, with just over 5% of visitors predicted to support this option. These findings provide further evidence suggesting that visitors would prefer to tolerate some amount of campsite sharing in order to ensure that the park does not build a large number of new campsites or require visitors to follow prescribed, fixed itineraries.

Descriptive and Prescriptive Components of Carrying Capacity

The research program described in this chapter illustrates the role of both descriptive and evaluative/prescriptive components of carrying capacity analysis and park and wilderness management more broadly. It also illustrates the ways in which research can help inform the process. In this case study, the descriptive component was an analysis of the relationship among potential indicators of the wilderness experience at Isle Royale. For example, to what degree would wilderness camping permits have to be reduced to achieve selected standards for campsite sharing? These relationships were determined using simulation modeling, and resulting data were used to design a series of practical and feasible management alternatives. These management alternatives were then used to help address the prescriptive component of carrying capacity. Which management indicators are most important to wilderness visitors? How would wilderness campers prefer to make tradeoffs among competing indicators? Which management alternatives should be implemented? A stated choice-based visitor survey was used to help answer these questions.

Study findings suggest that management actions, such as prescribed, fixed itineraries, large reductions in visitor use, and development of additional wilderness campsites are unfavorable to visitors to Isle Royale, despite the fact that they would substantially reduce campsite sharing. Further, the effect of campsite sharing on visitors' utility is relatively weak. These findings suggest that visitors would prefer to forfeit some degree of campsite solitude in order to ensure that their wilderness camping experiences are relatively free from management control and the environmental impacts of further development of the wilderness. Management of wilderness camping at Isle Royale will ultimately require management judgment about appropriate management practices. But these judgments should be as informed as possible about the potential effectiveness of these management actions and their consequences to the quality of the wilderness experience.

CHAPTER 12

Indicators and Standards at Boston Harbor Islands National Recreation Area

The optimum . . . is, then, less than the maximum.

Boston Harbor Islands National Recreation Area is a relatively new unit of the national park system. The park includes thirty-four diverse islands located in the inner and outer harbors of Boston. The islands are rich in natural and cultural history and offer a diverse array of recreation opportunities to residents of the Boston metropolitan area and beyond. Many of the islands are accessible through an inter-island ferry and shuttle-boat system. As the park has become better known and additional recreation facilities and services have been provided, visitor use has expanded to over a quarter million visits per year, and this is likely to grow substantially.

In conjunction with the *General Management Plan (GMP)* developed for this new park, an assessment of carrying capacity was also conducted, following the Visitor Experience and Resource Protection (VERP) framework outlined in chapter 2. An integrated program of natural and social science research was conducted to support this analysis (Manning et al. 2005a). Natural science research included an inventory and assessment of recreation-related impacts to park resources following methods developed in the recreation ecology literature (Cole 1989; Leung and Marion 2000; Marion 1991; Marion and Leung 2001). Social science research included surveys of park visitors using several of the research methods described in part 2 of this book. This program of research identified several potential indicators of resource and experiential conditions, including trail and campsite impacts, crowding, litter, and vandalism. Data were also gathered to help provide an empirical foundation for setting standards for these indicator variables.

Trail and Campsite Impacts

The program of natural science research found that recreation-related impacts to trails and campsites were evident in several areas of the park and included destruction of groundcover vegetation, soil erosion, campsite expansion, and trail widening. Data on the extent of current impacts offer some sense of what may be realistic standards for these indicator variables. However, visitor perceptions of these impacts offer an additional perspective. An initial survey of visitors found that many visitors noticed and objected to these recreation-related impacts. Using the visual research approach described in chapter 6, computer-edited photographs were prepared illustrating a range of potential trail and campsite impacts. Photographs for campsite impacts are shown in figure 12.1. These photographs were incorporated into a second visitor survey to measure normative standards for these indicator vari-

Figure 12.1. Study photographs of campsite impacts.

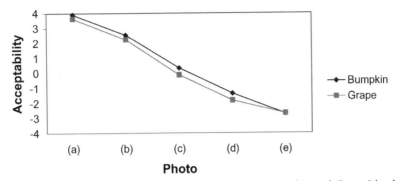

Figure 12.2. Social norm curves for campsite impacts at Bumpkin and Grape Islands.

ables, as described in chapter 5. The social norm curves for campsite impacts for Bumpkin and Grape islands (the only islands that allow camping) are shown in figure 12.2 and provide an empirical basis for setting experientially based standards for campsite impacts on these islands. For both islands, the level of campsite impacts depicted in study photograph (12.1[c]) represents the point at which average evaluations fall out of the acceptable range and into the unacceptable range.

Crowding

Findings from the initial visitor survey suggested that crowding is an emerging issue at several locations and is manifested in the number of people-at-one-time (PAOT) at attraction sites and the number of other visitors encountered while hiking on park trails. Normative standards for these indicator variables were addressed in the second visitor survey. Computer-edited photographs showing a range of PAOT at key attraction sites were prepared. The photographs for the historic lighthouse on Little Brewster Island are shown in figure 12.3. Visitors were asked to rate the acceptability of each of these photographs, and the resulting social norm curve for Little Brewster Island is shown in figure 12.4. Interestingly, respondents prefer to see some level of visitor use (up to around 55 PAOT). However, after this point, ratings of acceptability drop sharply and fall out of the range of acceptability at about 90 PAOT.

Trail-use levels at Boston Harbor Islands are generally not high enough to warrant use of a visual research approach to measuring associated normative standards. Therefore a numerical approach was used. Visitors were asked to report in an open-ended manner the maximum acceptable number of other visitors encountered per hour while hiking on park trails. Other evaluative dimensions of preference, management action, and displacement (as discussed in chapter 5) were also used in this series of questions. Findings for the five islands with

(a) 0 people  (b) 55 people

(c) 110 people  (d) 165 people

(e) 218

Figure 12.3. Study photographs for PAOT on Little Brewster Island.

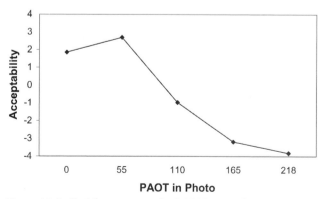

Figure 12.4. Social norm curve for PAOT on Little Brewster Island.

TABLE 12.1. Normative standards (means) for number
of other visitors encountered/hour along park trails at
Boston Harbor Islands National Recreation Area, MA

| Island | Evaluative dimension | | | |
	Preference	Acceptability	Management action	Displacement
Bumpkin	9.6	18.7	26.6	31.0
Grape	8.4	13.5	20.1	27.2
Lovells	18.1	27.5	26.6	28.2
Peddocks	15.6	26.3	24.1	27.3
Worlds End	10.4	19.2	26.6	32.9

established trail networks are shown in table 12.1. There were both similarities
and differences in encounter-related normative standards across the five islands.
For example, visitors to Grape Island preferred to see an average of 8.4 visitors
per hour along trails compared to an average of 18.1 for visitors to Lovells Island.
However, the displacement-based normative standard was quite similar for all
five islands ranging from 27.2 to 32.9 visitors per hour. Study data for both
PAOT at attraction sites and number of other hikers encountered per hour provide an empirical basis to set standards for these indicator variables.

Litter and Graffiti

The natural science-based inventory and assessment of recreation-related impacts
found that graffiti was an emerging problem at selected locations in the park. Similarly, findings from the initial visitor survey identified both litter and graffiti as
impacts that detract from the quality of the visitor experience. Thus, litter and graffiti may be good potential indicators of both resource and experiential conditions.

Normative standards for litter and graffiti were measured using visual
research methods and the Photometric Index (PI) approach developed by Keep
America Beautiful (KAB), a nonprofit group focused on the issue of litter (Keep
America Beautiful 2000; Budruk and Manning 2004; Budruk and Manning,
2006). In this approach, a standardized (16 x 6 ft.) horizontal grid of 96 cells created according to KAB specifications is overlaid on a park scene. Litter accumu-
lation is measured according to the number of cells occupied by litter. Each of
the 96 cells counts equally toward a PI rating of 0 to 96. If the same piece of litter covers multiple cells, each cell counts toward the scale value. In this study, the
grid created was overlaid on a park scene at Georges Island and four study photographs were prepared illustrating a range of litter, as shown in figure 12.5. The
four photographs used represented litter PI ratings of 0, 4, 8, and 12. These litter PI ratings were selected to represent a realistic range of litter in the park.

Figure 12.5. Study photographs for litter.

Respondents were asked to rate the acceptability of each photograph, and the resulting social norm curve for litter is shown in figure 12.6. These findings suggest there is little tolerance among visitors for litter.

Normative standards for graffiti were measured in a similar manner using the same PI approach. Four study photographs were prepared illustrating graffiti levels at Georges Island with PI ratings of 0, 26, 62, and 94, as shown in figure 12.7. Respondents were asked to rate the acceptability of each photograph, and the resulting social norm curve is shown in figure 12.8. As with litter, these findings suggest there is little tolerance among visitors for graffiti. Study data for litter and graffiti provide an empirical basis to set standards for these indicator variables.

Formulating Indicators and Standards

After completion of the natural and social science research components, a series of planning workshops was held to formulate indicators and standards for the park. Workshop participants included National Park Service staff, representatives

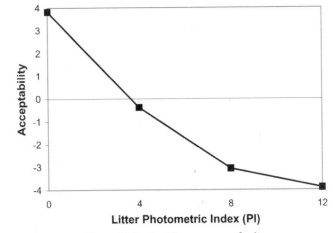

Figure 12.6. Social norm curve for litter.

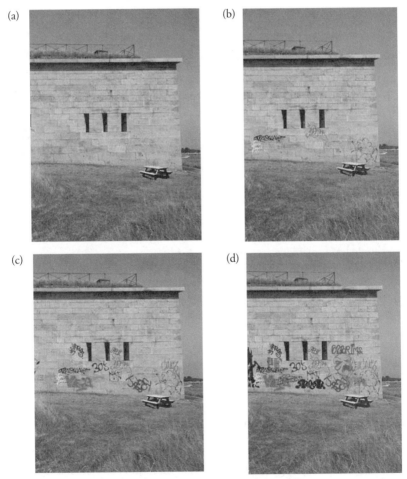

Figure 12.7. Study photographs for graffiti.

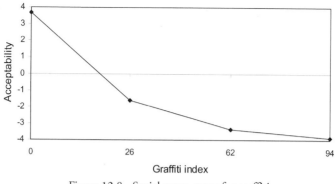

Figure 12.8. Social norm curve for graffiti.

from partner agencies and institutions, and research staff. The GMP for the park established six zones into which all parklands are assigned. The management objectives for each of the zones provided initial guidance concerning appropriate indicators and standards for resource and experiential conditions. Study findings from the natural and social science research provided a more empirical basis for formulating indicators and standards.

For example, management objectives developed in the park's GMP specify that the Natural Features Emphasis zone of the park should provide opportunities for visitor solitude, while the Visitor Services and Park Facilities Emphasis zone should provide opportunities for a more social experience. Study findings from the social science data outlined a range of potential standards for visitor density (from "preference" to "displacement"), including PAOT at attraction sites and encounters with other visitors along trails. For the Natural Features Emphasis zone (emphasizing solitude), a standard near the preference end of the range was selected (to ensure high levels of solitude), while a standard near the management action/displacement end of the range was selected (because solitude is not as important in this area) for the Visitor Services and Park Facilities Emphasis zone (emphasizing a social experience).

Findings from the natural science component of the study were also utilized in the formulation of standards. Study data documented the current level of recreation impacts in the park, and this information helped determine realistic standards that might be formulated. The scientific literature on related studies was also used. For example, research suggests that recreation-caused soil erosion at parks may range form 20% to 30% of mineral soil exposure (Kuss and Morgan 1984; Kuss and Morgan 1986). The park selected the lower end of the range (20%) as the standard for soil erosion based on this research and the fact that only a few sites in the park currently violate this standard.

A third example involves both natural and social science research findings. Natural science research found that visitors were causing impacts at campsites

TABLE 12.2. Selected indicators and standards for visitor carrying capacity at Boston Harbor Islands National Recreation Area, MA

Management zone and islands	People at one time (PAOT) (figure 12.4)	Tour group size	Trail encounters (hikers/hour)	Eroded tread on official trails (ft.^{2})	Official picnic sites and campsites	
					Campsite impacts (figure 12.2)	Bare soil exposure (%)
NATURAL FEATURES EMPHASIS (Button, Calf, Green, Hangman, Langlee, Little Calf, Middle Brewster, Outer Brewster, Raccoon, Ragged, Sarah, Shag Rocks, Sheep, Slate, Snake)						
MANAGED LANDSCAPE EMPHASIS (Bumpkin, Gallops, Grape, Great Brewster, Webb Memorial Park, Worlds End, Rainsford)			5–15	50	P	20
HISTORIC PRESERVATION EMPHASIS (Little Brewster, The Graves)	60[a]	30[a]				20
MULTIPLE MANAGEMENT EMPHASES (Deer, Georges, Long, Lovells, Moon, Nut, Peddocks, Spectacle, Thompson)	185–200[b]	30	15–25			20
SPECIAL-USE EMPHASIS (Nixes Mate)						

Note: P = applicable to public-use islands only; blank cell means indicator not recommended for zones or islands.
[a]Little Brewster only
[b]Georges and Long only

through destruction of groundcover vegetation and resulting sheet erosion of exposed bare soil. From an ecological standpoint, this process can lead to degradation of the park's natural condition. However, natural science data provides only a partial picture to suggest an appropriate standard for this indicator variable. Social science data can complement natural science data by suggesting a standard for campsite condition from an aesthetic standpoint based on visitors'

Litter (figure 12.6)	Graffiti (figure 12.8)	Density of social trails (ft./acre)	Area disturbed unofficial recreation sites (ft^2)	% visitors who have quality information	% visitors who have adequate information
1–2	2	10	0		
1–2	2	10	0	75	P
1–2	2[a]	10[a]	0[a]	75[a]	P
1–2	2	10–50	0	75	P
1–2	2				

perceptions. The social science findings described earlier for Bumpkin and Grape islands suggest the point at which campsite impacts are judged as unacceptable by visitors, and these data can be used to help set a socially acceptable standard for the ecological indicator of campsite impacts.

Workshop sessions resulted in sets of resource and social indicators and standards for each zone on each island. These indicators and standards are designed to meet the management objectives for each zone as defined in the GMP and to provide for an appropriate range of resource conditions and visitor opportunities throughout the park. Indicators and standards that were selected are summarized in table 12.2.

CHAPTER 13

Estimating Carrying Capacity
of Alcatraz Island

*Plainly, we must soon cease to treat the parks as commons
or they will be of no value to anyone.*

This chapter describes a program of research to help estimate and manage carrying capacity of Alcatraz Island. The island is a historic site within Golden Gate National Recreation Area, California, and is a heavily visited tourist attraction. It is located in San Francisco Bay and is widely known for its history as a federal prison for incorrigible criminals. This history has been romanticized and popularized in several books and movies. Consequently, demand to visit Alcatraz is high, and there is concern that visitation eventually may exceed carrying capacity. Visitors access the island by means of a ferry system, and the number of tickets sold is regulated by the National Park Service (NPS). The principal visitor attraction is the prison cellhouse, and most visitors take a cellhouse audio tour. A program of research was undertaken to help estimate and manage carrying capacity of the island (Manning et al. 2002a) and to help apply the Visitor Experience and Resource Protection (VERP) framework described in chapter 2.

Indicators and Standards

Two components of research were conducted. The first gathered data that would help formulate indicators and standards for the visitor experience at Alcatraz Island. A survey of a representative sample of visitors was conducted. The study questionnaire addressed both indicators and standards for the quality of the visitor experience. Indicators were addressed through a series of open- and close-ended questions, as described in chapter 4. Open-ended questions probed respondents for what they enjoyed most and least about their visit to Alcatraz: What did you enjoy most about your visit? What did you enjoy least about your visit? If you could ask the National Park Service to improve some things about the way visitors experience Alcatraz Island, what would you ask managers to do? Close-ended questions asked respondents to rate the seriousness of several potential problem issues including the number of visitors on the island, the number of visitors on the ferry, the number of visitors on the tour of the prison cellhouse, the number of visitors in the bookstores/gift shops, visitors making too much noise, and visitors not following rules and regulations. A three-point response scale was used that ranged from 1 ("not a problem") to 3 ("big problem").

Questions regarding standards focused on crowding-related issues. Specifically, a series of questions measured crowding-related normative standards for the prison cellhouse, the principal visitor attraction. Since use levels are relatively high in the prison cellhouse, a visual approach was used to measure crowding norms, as described in chapter 6. A series of computer-edited photographs was prepared for the Michigan Avenue corridor of the prison cellhouse (an integral part of the cellhouse tour) showing a range of visitor-use levels or people-at-one-time (PAOT). Study photographs are shown in figure 13.1. Respondents were asked to rate the acceptability of each photograph. In addition, respondents were asked to judge the photographs using two other evaluative dimensions—preference and management action—as described in chapter 5. Finally, respondents were asked "Which photograph looks most like the number of visitors you *typically saw* on your tour of the cellhouse today?"

Study findings yielded information on potential indicators and standards for the visitor experience at Alcatraz Island. Responses to open-ended questions were coded verbatim and then grouped into similar categories. Frequency distributions and mean values were calculated for responses to close-ended questions. Based on this analysis, the number of visitors in the prison cellhouse emerged as especially important in defining the quality of the visitor experience at Alcatraz. The vast majority of respondents (75%) reported that the audio tour of the cellhouse was the element of their experience that they enjoyed the most. Moreover, the number of visitors on the tour of the cellhouse was rated by respondents as the most problematic of six potential issues on the island; 54% reported that this was either a "big problem" or a "small problem." Thus, the number of visitors in the cellhouse is a good indicator in that it is measurable, manageable, and important in defining the quality of the visitor experience.

Data on standards for the number of people in the cellhouse are shown in figure 13.2 and table 13.1. Figure 13.2 graphs the average acceptability ratings for the six study photographs. As this social norm curve illustrates, 44 PAOT in Michigan Avenue represents a threshold of acceptability—the point at which

(a) 10 people

(b) 22 people

(c) 34 people

(d) 46 people

(e) 58 people

(f) 70 people

Figure 13.1. Study photographs of PAOT on Michigan Avenue in the prison cellhouse.

aggregate sample ratings fall out of the acceptable range and into the unacceptable range. Table 13.1 reports summary findings for all three evaluative dimensions explored. Respondents would prefer to see an average of 25 PAOT in Michigan Avenue, think the NPS should allow a maximum of 44 PAOT, and (as noted) would find a maximum of 44 PAOT to be acceptable. These data suggest a potential range of crowding-related standards. Neither of the points defining this range (25 or 44), nor any of the points along this range, are necessarily more

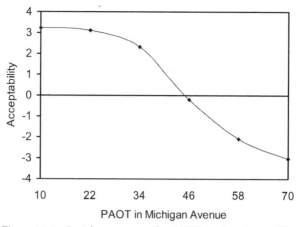

Figure 13.2. Social norm curve for PAOT in the prison cellhouse.

TABLE 13.1. Alternative crowding-related normative standards
for PAOT in the prison cellhouse, Alcatraz Island, CA

	Evaluative dimension	
Preference	Acceptability	Management action
25	44	44

Note: PAOT = people-at-one-time.

"valid" than any other. Each point has potential strengths and weaknesses. For example, standards based on preference-related norms may result in very high quality visitor experiences but would restrict access to a relatively low number of visitors. In contrast, standards based on acceptability or management action allow access to a greater number of visitors but may result in visitor experiences of lesser quality. Findings that offer insights into multiple evaluative dimensions provide a potentially rich base of information and may lead to formulation of the most thoughtful and informed standards. Such data allow more explicit understanding of the potential tradeoffs between use level and quality of the visitor experience.

Estimating Carrying Capacity

The second component of research focused on developing a computer-based simulation model of visitor use at Alcatraz Island. The primary purpose of the model was to determine the relationship between the number of visitors to the island and potential standards for the maximum PAOT in Michigan Avenue of the cellhouse. In this way, carrying capacity could be estimated in relationship to crowding-related standards for the cellhouse. The model was constructed using the commercial, object-oriented, dynamic simulation package Extend, as described in chapter 8. Model input was based on detailed visitor counts and observations, including number of visitors per ferry, frequency of ferries, length of time between debarkation of visitors on the island and their arrival into the cellhouse audio-tour ticket line, time spent in the cellhouse audio tour ticket line, and time spent touring the cellhouse. The model was designed to estimate PAOT in Michigan Avenue based on the number of visitors delivered to the island by the ferry system.

A short survey was also administered to a sample of visitors as they completed the cellhouse audio tour to determine the relationship between PAOT in Michigan Avenue, as reported by visitors using the computer-edited photographs, and the PAOT in Michigan Avenue, as estimated by the simulation model.

Once the simulation model was developed, it was used to estimate the maximum total daily-use levels (i.e., daily carrying capacities) that could be accommodated on the island without violating the normative crowding standards shown in table 13.1. Model output could be generated in several graphic and numerical forms. For example, figure 13.3 traces minute-by-minute PAOT levels in Michigan Avenue over the duration of a simulated day. This particular model run was generated using a current average summer day total-use level of 4,464 visitors (derived from the counts of visitor use taken to construct the model). It can be seen from the graph that the number of visitors in Michigan Avenue fluctuates between about fifty and ninety throughout most of the day. The model was ultimately run multiple times (to average out the randomness associated with each model run) to estimate the maximum total daily-use levels that could be accommodated on the island without violating each of the crowding-related norms shown in table 13.1 more than 10% of the time. It is clear from the findings shown in table 13.2 that there is a range of daily carrying capacities (from approximately 2,500 visitors per day to approximately 4,800

TABLE 13.2. Alternative daily carrying capacities of Alcatraz Island, CA

Crowding-related standards	Daily carrying capacity
Preference (25 PAOT)	2,560
Acceptability (44 PAOT)	4,800
Management action (44 PAOT)	4,800

Note: PAOT = people-at-one-time.

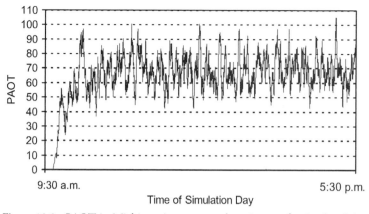

Figure 13.3. PAOT in Michigan Avenue over the minutes of a simulated day.

visitors per day) for Alcatraz Island, depending on the crowding-related standard selected. A daily carrying capacity can be implemented relatively easily through management of the ferry system serving the island.

As noted earlier, a short survey was also administered to visitors as they completed their audio tour of the cellhouse. Visitors were asked to select one of the study photographs of Michigan Avenue that looked most like the PAOT they typically saw at this location. The PAOT in the photographs selected by respondents averaged 35. The simulation model estimates that on the days when the survey was conducted there was an average of 70 PAOT in Michigan Avenue. This suggests there is a two-to-one ratio of the actual PAOT in Michigan Avenue and the number of people that can been seen at any one time. This is because people in the foreground and middle ground tend to obscure some people in the middle ground and background. Thus, the crowding-related norms derived from the study photographs underestimate the actual PAOT that can be accommodated in Michigan Avenue by about half. That is, if visitors report that a maximum of 44 PAOT is acceptable in Michigan Avenue (the PAOT visitors find acceptable to see based on the study photographs), then there actually can be approximately 88 PAOT in Michigan Avenue.

Given these study findings, what should the carrying capacity of Alcatraz Island be? What is an appropriate crowding-related standard, and how should this affect management of Alcatraz? It seems clear that the number of visitors in the prison cellhouse is a potentially good indicator of the quality of the visitor experience. Nearly all visitors take the audio tour of the prison cellhouse and feel this is the highlight of their visit. However, there is evidence that visitors are concerned with growing use levels in the cellhouse. Visitors rated crowding in the cellhouse as the most problematic of several visitor-related issues. Moreover,

visitor perceptions of current use levels in the cellhouse are approaching the maximum PAOT judged acceptable. Finally, visitors rated the prison cellhouse as
"somewhat crowded" (an average of 4.2 on a 9-point crowding scale that ranged
from 1 ["not at all crowded"] to 9 ["extremely crowded"]), and this represented
the highest level of crowding for any location on the island. However, standards
for this indicator vary substantially, depending upon the evaluative dimension
used in the study (i.e., the type of visitor experience). Crowding norms ranged
from a low associated with "preference" (a very high-quality experience) to a high
associated with "acceptability" and "management action" (a lower-quality experience). The carrying capacity of the prison cellhouse, and ultimately Alcatraz
Island as a whole, varies accordingly.

As discussed in part 1 of this book, carrying capacity requires some element
of management judgment. However, the data developed in this research program offer empirical and rational foundations for such judgments. Study data
suggest PAOT standards in the prison cellhouse in the range of 25 to 44,
depending on the evaluative dimension used. Additional study data and other
considerations may suggest a standard and associated carrying capacity in the
high end of this range.

First, the low end of this range is associated with visitor "preferences" and
does not include explicit consideration by respondents of the tradeoffs between
crowding (or *lack* of crowding) and maintaining reasonable public access to Alcatraz. In contrast, the upper end of this range is associated with the use level visitors feel the NPS "should allow" and is more explicitly informed by tradeoffs
between crowding and public access. (As described in chapter 5, the question
asked visitors to select the photograph that represented the point at which visitors should "be restricted from touring the cellhouse.") This issue is reinforced
by two other sets of questions that addressed tradeoffs between crowding and
access. Visitors were asked to rate the importance (on a five-point scale ranging
from 1 ["very important"] to 5 ["very unimportant"]) of crowding (defined as
the "ability to visit Alcatraz Island without it being crowded") and availability of
tickets to visit the island (defined as the "ability to get a ticket when wanted").
These attributes were judged to be nearly equally important (average importance
ratings of 1.76 and 1.82, respectively). Visitors were also asked to allocated 10
'points of importance' between crowding and access. Once again, both issues
were judged to be important, with crowding receiving an average of 5.4 "points
of importance," and access receiving an average of 4.6.

Second, public demand to visit Alcatraz Island is very high, and it may be
unrealistic and ultimately unacceptable to greatly restrict public access to achieve
a very low level of crowding.

Third, the crowding norms measured in this study are slightly under-
estimated. A small number of respondents reported that all of the study photographs were acceptable or that the NPS should not limit use at any point

represented in the study photographs. These responses could not be included mathematically in the values reported in table 13.1.

Fourth, current use levels are perceived by respondents to be about 41 PAOT in Michigan Avenue, and visitors reported feeling "somewhat crowded." This suggests that use levels could be at least a little higher without causing very high levels of perceived crowding.

Fifth, since current use levels average near the high end of the range of potential standards, it may be politically unrealistic to set standards near the low end of the range.

Sixth, most visitors to Alcatraz are first-time visitors, and many live outside California and even outside the United States. This suggests that many visitors have few realistic options to visit Alcatraz, and management decisions about crowding should be made in light of strong public demand and limited opportunities to visit the island.

Finally, overall visitor enjoyment of Alcatraz Island is very high despite some degree of crowding in the prison cellhouse. For example, using a response scale from 1 ("strongly agree") to 5 ("strongly disagree"), visitors were asked the extent to which they agreed or disagreed with the statement: "I enjoyed my visit to Alcatraz Island." The average score was 1.52. This suggests that management decisions about carrying capacity of Alcatraz Island should not be made exclusively on the basis of crowding in the prison cellhouse.

As noted earlier in this book, it may be wise to consider a range of standards and associated carrying capacities (or experience "zones") for parks or sites within parks. Since Alcatraz Island is relatively small and the prison cellhouse is the icon visitor attraction, it may not be feasible to do this in a conventional spatial or geographic way. However, alternative standards and associated carrying capacities could be established for the summer peak-visitor-use season and a lower carrying capacity during the off seasons. Moreover, the NPS conducts special evening tours of the island, and these could be managed to a different standard, thereby providing a diversity of visitor experiences. Under this type of zoning system, visitors who are less tolerant of crowding would have an option to find the type of opportunity that is more in keeping with their normative standards, while visitors who are more tolerant of crowding would have a greater chance of visiting the island.

Managing Carrying Capacity

A final analytical approach used in this research employed the simulation model of visitor use to explore the effect of alternative ferry schedules on carrying capacity. Currently, ferries depart San Francisco for Alcatraz Island every half hour from 9:30 a.m. to 5:30 p.m. If ferry departures were reduced to once every hour, the simulation model estimates that daily carrying capacity would also be substantially reduced. Relatively large numbers of visitors arriving at the same time would result in many visitors seeing relatively large PAOT's in Michigan Avenue. For example (as reported in table 13.2), using the acceptability and management action standards of 44 PAOT, the simulation model estimates that approximately 4,800 visitors per day could be accommodated on Alcatraz Island without violating that standard more than 10% of the time. This analysis is based on the existing ferry schedule of departures every half hour. However, when ferries depart only once every hour, the simulation model estimates that only 3,200 visitors per day could be accommodated without violating the 44 PAOT standard more than 10% of the time. Similarly, for the preference-based standard, daily carrying capacity drops from approximately 2,500 visitors per day under the existing ferry schedule to approximately 1,840 visitors per day under the reduced ferry schedule. Comparable increases in carrying capacity are not possible in this case by increasing the frequency of ferry service beyond the existing schedule. For example, increasing the frequency of departures to every fifteen minutes would increase acceptability and management action-based carrying capacity from approximately 4,800 visitors per day (under existing ferry service) to approximately 4,896 visitors per day and would increase preference-based carrying capacity from approximately 2,560 visitors per day to approximately 2,656 visitors per day.

CHAPTER 14

Defining and Managing the Quality of the Visitor Experience at Muir Woods National Monument

The parks . . . are limited in extent . . . whereas population seems to grow without limit.

Muir Woods National Monument is located just north of San Francisco, California. It was established as a unit of the national park system in 1908 to preserve an impressive stand of thousand-year-old coast redwood trees. The park is small by national park standards (just over five hundred acres), but it is visited very intensively, accommodating over a million visits annually. The park offers six miles of trails, and visitors are required to stay on trails to help protect fragile soils, vegetation, and other resources. The park's main trail network, along Redwood Creek on the floor of the canyon, is hardened with paving or wooden boardwalks.

Indicators of the Quality of the Visitor Experience

As with many intensively visited parks, there is concern about managing Muir Woods within a carrying capacity. Since visitor use is restricted to a network of hardened trails, much of the concern has been focused on social carrying capacity. Under guidance of the Visitor Experience and Resource Protection (VERP) framework, as described in chapter 2, a program of research was designed and conducted to define and manage the quality of the visitor experience. The objectives of the research were to help identify indicators and standards for the quality of the visitor experience. Both open- and close-ended questions were used, as described in chapter 4. The survey was administered to a representative sample of visitors as they completed their visit. Open-ended questions asked visitors what they enjoyed most and least about their visit, what they would ask the National Park Service (NPS) to change about the way Muir Woods is managed, and the qualities and values of Muir Woods they thought were most important. Respondents who had visited the park previously also were asked what they thought had changed about Muir Woods, for the better and for the worse. Close-ended questions presented a list of twenty-one potential problems in the park. Respondents were asked to indicate whether they thought each item was "no problem," a "small problem," or a "big problem." Study findings for the close-ended questions are shown in table 14.1.

TABLE 14.1. Importance of potential problems
at Muir Woods National Monument, CA

Potential problem	N	Not a problem (1)	Small problem (2)	Big problem (3)	Don't know	Mean
			Percentage			
Too many people along trails	399	34.8	47.6	16.8	0.8	1.8
Large groups of people along trails	397	40.1	40.6	17.6	1.8	1.8
Too many people at the gift shop	396	48.5	18.9	7.1	25.5	1.4
Too many people at the café	394	48.0	13.7	4.1	34.3	1.3
Finding a parking place	400	25.3	30.0	42.8	2.8	2.2
Finding your way to Muir Woods	391	85.9	10.5	1.0=	2.6	1.1
Finding your way around Muir Woods	401	90.0	8.7	0.7	0.5	1.1
Visitors making too much noise	402	49.0	34.6	15.7	0.7	1.7
Too much noise from outside the park	397	88.4	7.3	0.3	4.0	1.1
People not obeying rules and regulations (e.g., walking off the trail, smoking)	398	68.6	18.8	6.8	5.8	1.3

Potential problem	N	Not a problem (1)	Small problem (2)	Big problem (3)	Don't know	Mean
			Percentage			
Lack of supervision of school/youth groups	397	59.2	8.8	2.3	29.7	1.2
Lack of printed materials about Muir Woods	399	66.9	23.1	5.5	4.5	1.4
Lack of educational talks and programs about Muir Woods	401	56.4	20.2	4.2	19.2	1.4
Lack of educational signs along trails	396	60.4	30.8	7.1	1.8	1.5
Waiting to use restrooms	398	70.4	13.8	3.5	12.3	1.2
Waiting to pay entrance fee	400	88.0	6.5	0	5.5	1.1
Unavailability of park rangers	397	58.2	19.4	3.5	18.9	1.3
Trails too highly developed	399	79.2	13.3	4.8	2.8	1.2
Too many facilities	401	86.3	7.5	1.2	5.0	1.1
Poor condition of trails	402	91.0	6.7	0.2	2.0	1.1
Litter along trails	401	89.0	7.5	1.2	2.2	1.1

Note: N = number of respondents.

In keeping with the definition of indicators as described in chapters 2 and 3, analysis of study data focused on identifying measurable, manageable variables that are affected by visitor-use levels and/or behavior and that are important in influencing the quality of the visitor experience. Four indicators were identified, including the number of visitors on park trails, large groups (primarily commercial tour and educational groups), difficulty finding a parking place, and visitor-caused noise (which disturbed the silence of the old growth forest and masked the natural soundscape).

Standards of Quality of the Visitor Experience

A second visitor survey explored standards for the indicator variables. Normative theory and methods were used as described in chapter 5.

Crowding

Since visitor-use levels are quite high at Muir Woods, a visual research approach (as described in chapter 6) was taken to measure normative standards for crowding on trails. Two sets of computer-edited study photographs were prepared illustrating a range of persons-per-viewscape (PPV) on the two types of trails in the park. The first set of photographs, shown in figure 14.1, addressed the main park trails along the floor of the canyon. These trails are very heavily used and are

(a) 0 PPV   (b) 6 PPV

(c) 12 PPV   (d) 18 PPV

(e) 24 PPV   (f) 30 PPV

Figure 14.1. Study photographs for PPV on the main trails.

(a) 0 PPV   (b) 2 PPV   (c) 4 PPV

(d) 6 PPV   (e) 8 PPV   (f) 10 PPV

Figure 14.2. Study photographs for PPV on the secondary trails.

hardened by wooden boardwalks. The second set of study photographs, shown in figure 14.2, addressed the lesser-used trails that branch off the main trails and climb the slopes of the canyon walls.

Respondents were asked to rate the acceptability of the photographs based on the number of visitors shown and were also asked to select the photograph that best represented the other evaluative dimensions of preference, management action, and displacement, as described in chapter 5. Finally, respondents were asked to select the photograph that best represented the level of use during their visit. Study findings for the main trails are shown in figure 14.3 and table 14.2, and findings for the secondary trails are shown in figure 14.4 and table 14.3.

These data suggest a range of potential crowding-related standards, but they also

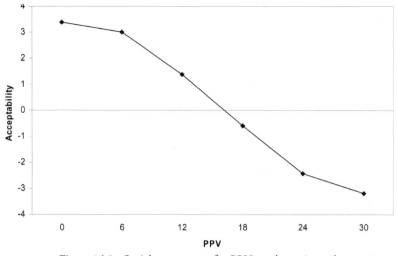

Figure 14.3. Social norm curve for PPV on the main trails.

TABLE 14.2. Normative standards for PPV on the main trails
at Muir Woods National Monument, CA

	Photo number (Mean PPV)	Photo number (Median PPV)
Acceptability	16.2 (PPV)	
Preference	2.1 (6.6)	2.0 (6.0)
Displacement	5.1 (24.6)	5.0 (24.0)
Management action	4.1 (18.6)	4.0 (18.0)
Typically seen	2.7 (10.2)	3.0 (12.0)

Note: PPV = persons/viewscape.

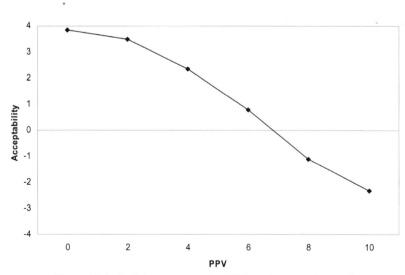

Figure 14.4. Social norm curve for PPV on the secondary trails.

TABLE 14.3. Normative standards for PPV on the secondary trails
at Muir Woods National Monument, CA

	Photo number (Mean PPV)	Photo number (Median PPV)
Acceptability	6.8 (PPV)	
Preference	2.0 (2.0)	2.0 (2.0)
Displacement	5.2 (8.4)	5.0 (8.0)
Management action	4.2 (6.4)	4.0 (6.0)
Typically seen	2.4 (2.8)	2.0 (2.0)

Note: PPV = persons/viewscape.

TABLE 14.4. Perceived crowding at Muir Woods National Monument, CA

Location	N	Not at all crowded 1	2	3	4	5	6	7	8	Extremely crowded 9	Mean
		Percentage									
Along the trails	385	22.9	24.4	22.9	10.9	6.5	7.8	3.4	1.0	0.3	3.0
In the parking area	385	13.8	11.4	11.7	7.0	5.2	12.2	8.1	15.1	15.6	5.2
Along the road to the park	387	21.4	26.4	15.0	9.0	4.9	9.0	6.7	4.7	2.8	3.4
In the visitor center	335	17.6	21.2	18.5	13.7	8.4	12.5	2.7	3.9	1.5	3.5

Note: N = Number of visitors

indicate that current use levels are relatively low in comparison to these standards. For example, visitors on the main trails find a maximum of 16.2 PPV to be acceptable, and currently see 10.2 PPV. Similarly, visitors on the secondary trails find a maximum of 6.8 PPV to be acceptable, and currently see 2.8 PPV. These findings are generally corroborated by another series of questions that asked respondents to report their level of perceived crowding in several locations in the park, as shown in table 14.4. Respondents reported relatively low levels of crowding on the park trails.

Parking

Findings in table 14.4 suggest that many visitors feel especially crowded in the parking areas, and this reinforces the potential importance of parking as an indicator variable. Normative standards for indicator variables related to parking were

TABLE 14.5. Maximum acceptable waiting time to find a parking place at Muir Woods National Monument, CA

Minutes	% of respondents
0	2.3
1	1.5
2	3.0
3	4.9
5	28.8
6	0.8
7	0.8
8	0.4
10	37.9
15	14.0
20	4.2
25	0.4
30 or more	1.2

Note: Mean = 9.2 minutes; Median = 10.0 minutes.

TABLE 14.6. Maximum acceptable walking time from parking place to park entrance at Muir Woods National Monument, CA

Minutes	% of respondents
1	1.1
2	2.2
3	1.1
4	0.7
5	23.7
6	0.4
7	1.8
8	1.1
10	40.3
12	0.4
15	15.8
18	0.4
20	7.6
25	0.7
30	1.1
40	0.4
60 or more	1.5

Note: Mean = 11.2 minutes; Median = 10.0 minutes.

measured using a numerical approach, as described in chapters 5 and 6. Two dimensions of parking were addressed: waiting time to find a parking place, and walking time from the parking place to the park entrance. (Because of the park's relatively small size, two small parking lots are provided. When these lots are full, visitors park on the shoulder of the road leading to the park.) Visitors were asked to report the maximum acceptable waiting time to find a parking place and the maximum acceptable walking time from the parking place to the park entrance. Study findings are shown in tables 14.5 and 14.6 and provide an empirical basis for selecting parking-related standards. Visitor reported average maximum acceptable times of approximately ten minutes for both indicator variables.

Group Size

Normative standards for group size were also measured using a numerical approach. Respondents were asked to rate the acceptability of group sizes that ranged from five to fifty in increments of five. The social norm curve derived

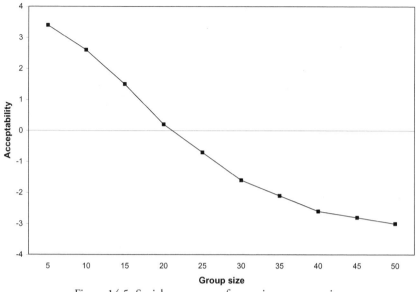

Figure 14.5. Social norm curve for maximum group size.

from study findings is shown in figure 14.5. It is clear that most visitors prefer relatively small group sizes, and that aggregate evaluations of group size fall out of the acceptable range and into the unacceptable range at about twenty.

Visitor-Caused Noise

Normative standards for noise were addressed in two phases of research. The first involved administration of a "listening exercise" to a sample of park visitors (Pilcher et al. 2006). Visitors were asked to sit at selected locations in the park and, using a checklist provided, record the types of natural and human-caused sounds they heard. For each type of sound heard, respondents were asked to rate on a nine-point scale how pleasing (+4) or annoying (–4) they found these sounds to be.

Study findings are summarized in figure 14.6, which plots the percentage of respondents who heard each type of sound and how pleasing or annoying that sound was. This figure is analogous to the importance-performance framework described in chapter 4. Sounds heard by large percentages of visitors and that are evaluated as highly annoying are good potential indicators, because they can be important in influencing the quality of the visitor experience. Likewise, sounds heard by a large percentage of visitors and that are evaluated as highly pleasing are also good potential indicators. Data from figure 14.6 suggest that sounds

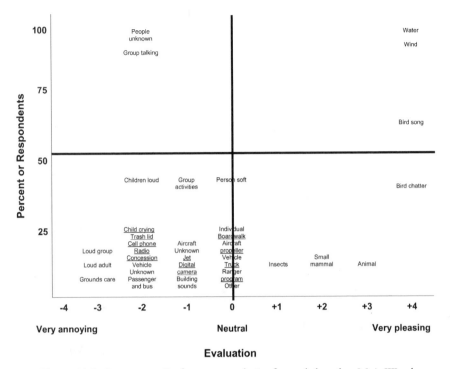

Figure 14.6. Importance-Performance analysis of sounds heard at Muir Woods.

constituting the former group include visitor-caused noise, such as strangers talk-ing; conversations within large groups; and loud children. Sounds constituting the latter group include wind blowing through the trees, water rushing in Red-wood Creek, and bird songs.

These findings were used to inform a second phase of research that meas-ured visitor-based normative standards for important characteristics of the park's soundscape (Manning et al. 2006). Using sound measurements taken in the park, a series of five thirty-second recordings was developed. The first recording represented the park's natural soundscape of wind, water, and bird songs. The next four recordings overlaid tracks of increasing levels of visitor-caused sounds, including talking and boisterous behavior. These five record-ings were incorporated into the visitor survey by asking respondents to listen to each recording and evaluate its acceptability. The resulting social norm curve for visitor-caused noise is shown in figure 14.7 and shows that aggre-gate judgments for the sample as a whole fall out of the acceptable range and into the unacceptable range at a visitor-caused noise level of thirty-seven decibels.

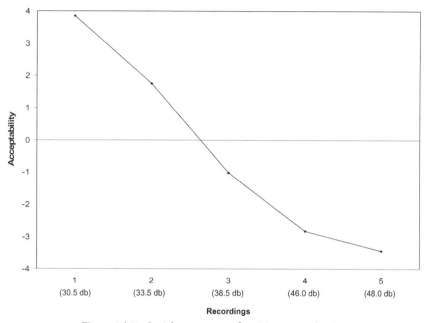

Figure 14.7. Social norm curve for visitor-caused noise.

Carrying Capacity Based on the Quality of the Visitor Experience

Findings from the program of research help provide an empirical basis for formulating indicators and standards of the quality of the visitor experience. Using the carrying capacity frameworks (such as VERP) described in chapter 2, these indicators and standards can be used to guide carrying capacity management. The indicators ultimately selected for the park will be monitored, and management actions implemented to ensure that the standards formulated are maintained.

As described in chapter 8, a more proactive approach to monitoring and managing carrying capacity can be used for crowding-related indicators. Since crowding on trails was a potentially important indicator of the quality of the visitor experience, a simulation model of visitor use of the park's trail network was developed. As a part of the initial visitor survey, a representative sample of visitors were given a trail map as they entered the park and asked to record where they hiked, how long it took them, and where they stopped and for how long. Counts of daily visitor-use levels and arrival rates were recorded by observers. These data were used to construct the model, and the model was designed to estimate PPV levels on the main and secondary trails based on total daily park

TABLE 14.7. Percentage of time that acceptability-based PPV standard is violated during peak-use hours at Muir Woods National Monument, CA

Multiple of current park use	Main trails %	Secondary trails %
1	0.1	0.0
2	7.1	0.4

Note: PPV = persons/viewscape: peak-use hours = 10:00 a.m. to 2:00 p.m.

use. Sample model output is shown in table 14.7. Column one is the multiple of the current average daily park-use level where *1* equals the current average daily use level, *2* equals double the current average daily use level, and so on. Columns two and three show the model estimates for the percentage of time during peak-use hours (defined as 10:00 a.m. to 2:00 p.m.) that PPV levels violate the acceptability standards, as shown in tables 14.2 and 14.3. Under current use levels, the PPV levels very rarely violate acceptability-based crowding standards (only about 0.1% of the time for both primary and secondary trails). The model estimates that park-use levels could double without violating acceptability-based PPV standards more than 10% of the time during peak-use hours. In this way, the simulation model can be used to monitor PPV levels (based on daily park attendance figures) and can estimate the maximum acceptable number of daily visitors without violating crowding-related standards. This type of information is important in formulating crowding-related standards, as well as managing the park to ensure that standards are maintained. For example, estimates of maximum acceptable daily park-use levels, including how this use should be spread over the hours of the day, can help inform decisions about sizing of parking lots, development and operation of a public transit system, and contracts for commercial tour and educational groups.

CHAPTER 15

Wilderness Management
at Zion National Park

Finally . . . comes the day of reckoning.

Zion National Park is located in southwestern Utah. It comprises nearly one hundred fifty thousand acres of sculptured canyons and soaring cliffs and offers some of the finest scenery in the desert Southwest. A *General Management Plan (GMP)* for the park was recently completed requiring that a complementary wilderness plan be prepared. Most of the park (approximately 90%) has been proposed for wilderness designation by the National Park Service (NPS), and use of this portion of the park has increased dramatically in recent years. For example, a permit is required for camping in wilderness areas and for day use of several popular canyons, and the number of permits issued increased by 97% over a recent five-year period.

The NPS is using the Visitor Experience and Resource Protection (VERP) framework (described in chapter 2) to guide their planning effort. A program of research was designed and implemented to support this planning. The primary focus of research was to provide an empirical basis to help formulate indicators and standards for resource, social, and managerial conditions of the wilderness portion of the park.

Wilderness Indicators

Several research methods that spanned the range of quantitative to qualitative (as outlined in chapter 4) were used to explore potential indicator variables for the wilderness portion of the park. Surveys were administered to several types of wilderness visitors, including multiday camping groups, day hikers in areas that did not require a wilderness permit, and day hikers in areas that did require a wilderness permit. Three types of surveys were administered: (1) a series of closed-ended questions that asked respondents to rate the importance of several potential indicators and the degree to which these indicators were attained or experienced; (2) a series of open-ended questions that asked respondents to report what they enjoyed most and least about their visit, what changes in management they would recommend, and (for those who had visited the park previously) what had changed about the park for the better or for the worse; and (3) a semistructured interview consisting of a series of in-depth, probing questions about the park and the associated wilderness experience.

Data from the closed-ended questions were organized into an importance-performance framework (as described in chapter 4), and findings for day-use hikers who did not need a permit are shown in figure 15.1. It is striking that all nine of the potential indicators included in the survey fall within the "keep up the good work" quadrant of the graph. For example, the opportunity to experience solitude while hiking was rated as "important" by the sample as a whole,

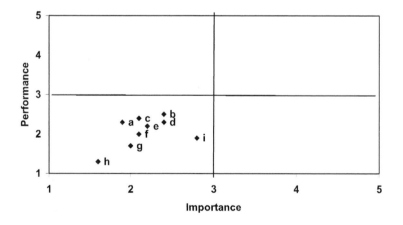

a. Opportunity to experience solitude while hiking
b. Opportunity to see few other visitors while hiking
c. Opportunity to avoid seeing large groups of visitors while hiking (groups greater than twelve visitors)
d. Opportunity to avoid seeing organized groups of visitors while hiking (e.g., clubs, scouts)
e. Opportunity to avoid seeing commercial groups of visitors while hiking (i.e., groups where visitors pay to participate)
f. Opportunity to use trails that do not show a lot of recreation-related impact
g. Opportunity to avoid seeing and/or hearing aircraft
h. Opportunity to use the park's shuttle bus system to get to and from trailheads
i. Opportunity to avoid seeing evidence of recent fires (either naturally ignited fires or management prescribed burns)

Figure 15.1. Importance/performance graph for Zion National Park day use that does not require a permit.

and the park was rated as "good" with respect to providing solitude while hiking. Findings from this component of the study suggest that solitude and the absence of trail impacts are potentially good recreation-related indicators. These findings were similar across all types of wilderness areas and visitors.

Responses from the open-ended questions were grouped into categories. Three categories of issues emerged as potential indicators of wilderness conditions and associated experiences for all types of wilderness areas and visitors: (1) absence of recreation-related impacts to wilderness resources, (2) solitude/lack of crowding, and (3) peacefulness/quiet.

Findings from the semistructured interviews offered insights into potential indicators (Freimund et al. 2004). For example, many respondents once again emphasized the importance of solitude to the quality of the wilderness experience. Yet, other quantitative components of the research program (described later) found that many day-use hikers, particularly those hiking to areas where no permit is required, saw dozens, even hundreds, of other hikers. Additional findings from the interviews suggest that many respondents saw large numbers of other visitors only at selected locations (for example, at trailheads and attraction sites) and that they were able to achieve adequate solitude during other portions of the hike. Other respondents "coped" with high-use trails by hiking to alternative locations and/or hiking during off-peak periods (e.g., on weekdays rather than weekends, early or late in the day).

Day-Use Wilderness Standards

Surveys of several types of wilderness users were administered to help formulate standards for selected indicator variables, including encounters with other groups along trails, trail and campsite impacts, trail and campsite development, and group size. This section describes several examples taken from different locations within the wilderness portions of the park that are used by day visitors.

The Grotto and Weeping Rock areas of the park provide access to trails that are heavily used by day visitors and do not require wilderness permits for day use. Representative samples of day users at these locations were administered a questionnaire that addressed normative standards for encountering other hikers per day along these trails. Respondents were asked to rate the acceptability of seeing 0, 20, 40, 60, 80, and 100 hikers per day. Resulting social norm curves for both areas are plotted in figure 15.2. The social norm curves are quite similar, though hikers using the Grotto area are somewhat more tolerant of other visitors than are hikers using the Weeping Rock area: the norm curve crosses the neutral point of the acceptability scale at approximately 74 visitors encountered for the former group and at approximately 60 encounters for the latter. Nevertheless, both groups report normative standards that are relatively high, especially for wilderness areas. However, these areas are served by trailheads that are directly on the

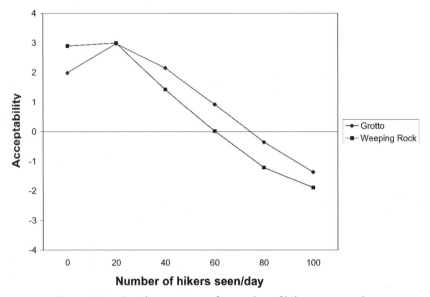

Figure 15.2. Social norm curves for number of hikers seen per day
along trails in the Grotto and Weeping Rock areas.

route of the park's mandatory shuttle-bus system that lead to especially popular hiking destinations and that are heavily used by day hikers.

Survey respondents were also asked to report the approximate number of visitors encountered during their hike and how crowded respondents felt along the trail. Hikers using the Weeping Rock area reported seeing an average of approximately 25 visitors during their hike (compared to 60, which was their threshold of acceptability). Respondents also reported very low levels of crowding, averaging 1.9 on a nine-point scale that ranged from 1 ("not at all crowded") to 9 ("extremely crowded"). Respondents using the Grotto area reported seeing nearly twice as many visitors (an average of approximately 47) and also reported a higher level of crowding (averaging 3.4 on the nine-point response scale).

Day-use visitors to the Narrows portion of the park's wilderness were administered a similar questionnaire. This area does not require a wilderness permit for day use. The Narrows offers an unusual hiking opportunity defined by a very narrow and deep "slot canyon" formed by the Virgin River. Throughout much of this hike, visitors must walk directly in the river, even swimming through deeper sections. Because of the special character of this area, including its national and international reputation, it is very heavily used. This suggested that a visual research approach (as described in chapter 6) would be useful in measuring crowding-related norms. A series of seven computer-edited photographs were prepared illustrating a range of use levels for a typical section of the canyon, as shown in figure 15.3. A

(c) 12 people

(d) 18 people

Figure 15.3. Study photographs illustrating a range of use levels in the Narrows.

(e) 24 people

(f) 30 people

(g) 36 people

Figure 15.3. *Continued*

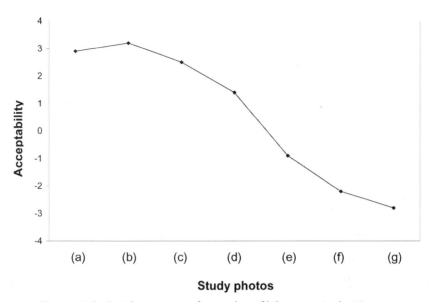

Figure 15.4. Social norm curve for number of hikers seen in the Narrows.

representative sample of visitors was asked to rate the acceptability of each of the study photographs, and the resulting social norm curve is shown in figure 15.4. These data indicate that aggregate ratings of the sample fall out of the acceptable range and into the unacceptable range when approximately 22 other hikers can be seen at any one time.

Respondents were also asked to indicate the photograph that looked most like the number of hikers typically seen and how crowded respondents felt while hiking. The average number of hikers seen as shown in the study photographs was approximately 16, and respondents reported feeling moderately crowded, averaging a score of 3.8 on the nine-point response scale described earlier.

Day-use visitors at all of the three study locations were asked to report normative standards for size of hiking groups. Respondents were asked to rate the acceptability of the following group sizes: 4, 6, 8, 10, 12, 14, and 16. The resulting social norm curve is shown in figure 15.5 and suggests that hiking groups larger than 10 are generally considered to be unacceptable by day hikers. However, norm intensity or salience is not high (as defined by the amplitude of the curve; this issue is discussed in chapter 5), suggesting that group size may not be an especially important indicator variable for day-use hiking.

Finally, day-use visitors to areas of the park that require wilderness permits were also administered questionnaires exploring standards for selected indicator

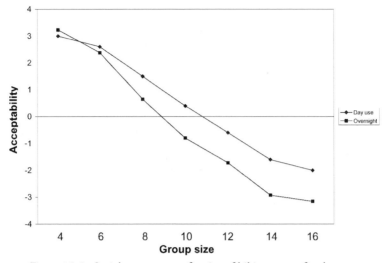

Figure 15.5. Social norm curves for size of hiking group for day use
and overnight hikers at Zion National Park.

variables. Most of these areas are narrow slot canyons that have become popular
for technical "canyoneering." For example, a representative sample of these visitors was asked to rate the acceptability of seeing and/or hearing a range of 0 to
16 other groups per day while in these canyons. The resulting social norm curve
is shown in figure 15.6. Visitors to these areas are clearly less tolerant of encoun-

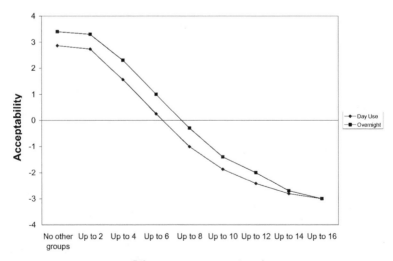

Figure 15.6. Social norm curves for groups encountered per day in areas requiring a
wilderness use permit for day use and for overnight wilderness visitors.

tering other users than are visitors to areas that do not require wilderness permits (as shown in figure 15.2). Normative standards for encounters with other groups for day-use visitors to areas requiring wilderness permits are similar to those of overnight wilderness visitors.

Overnight Wilderness-Use Standards

All overnight visitors to the wilderness portion of the park are required to obtain a wilderness-use permit. A representative sample of these visitors was administered a questionnaire to determine their normative standards for several potential indicator variables. As described earlier, two of these indicators were group size and number of visitors encountered per day along trails. These standards were measured for both day and overnight visitors, and resulting social norm curves are shown in figures 15.5 (group size) and 15.6 (number of groups encountered). Overnight wilderness visitors are slightly less tolerant of large groups than are day visitors, but the former have a stronger norm intensity than the latter. This suggests that group size is a better or more salient indicator for overnight wilderness visitors than for day visitors. Overnight visitors and day-use visitors to areas that require a wilderness-use permit have very similar normative standards for encounters with other groups along trails, and both social norm curves suggest relatively high norm intensity. Both social norm curves fall out of the acceptable range and into the unacceptable range at around six to eight groups encountered per day.

Normative standards of overnight visitors were also measured for trail and campsite impacts and associated development designed to mitigate such impacts. These potential indicators lend themselves to the visual approach described in chapter 6, and resulting computer-edited study photographs are shown in figures 15.7 through 15.10. In each case, respondents were asked to rate the acceptability of all study photographs, and resulting social norm curves are shown in figures 15.11 through 15.14. These findings suggest that visitors are relatively intolerant of visitor-caused impacts to trails but considerably more tolerant of visitor-caused impacts to campsites. Moreover, visitors find most of the trail and campsite development practices illustrated in the study photographs to be relatively acceptable.

Visitor Access versus Resource Protection

An additional component of the research at Zion National Park was designed to explore tradeoffs that wilderness visitors were willing to make between access to the park and protecting park resources. This component of the study used stated choice analysis as described in chapter 7. A representative sample of wilderness

Figure 15.7. Study photographs illustrating a range of trail impacts.

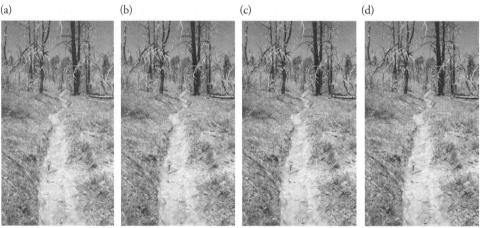

Figure 15.8. Study photographs illustrating a range of trail development.

Figure 15.9. Study photographs illustrating a range of campsite impacts.

Figure 15.10. Study photographs illustrating a range of campsite
development at Zion National Park.

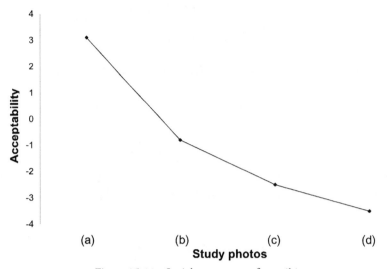

Figure 15.11. Social norm curve for trail impacts.

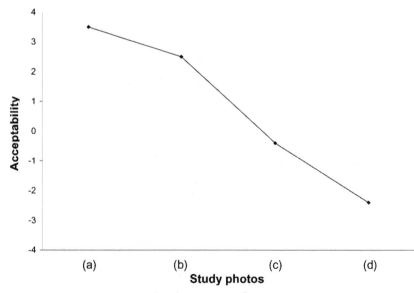

Figure 15.12. Social norm curve for campsite impacts.

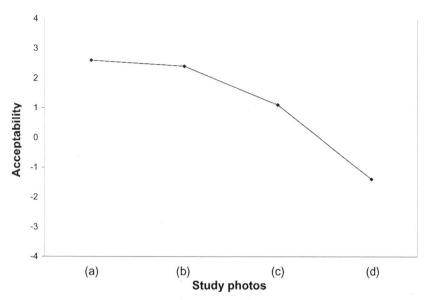

Figure 15.13. Social norm curve for trail development.

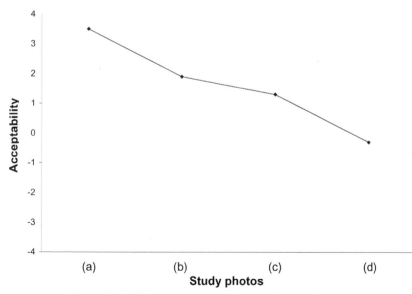

Figure 15.14. Social norm curve for campsite development.

visitors was asked a series of questions comprising paired comparisons. Each paired comparison showed two of the five study photographs illustrating a range of trail impacts (as shown in figure 15.7). Respondents were asked to pick the photograph representing the trail conditions they preferred. Respondents were then asked if they would be willing to accept less than a 100% chance of receiving a wilderness-use permit to hike the trail they had just used in order to ensure that the trail impact conditions they preferred would be maintained. Percentages of 10, 25, 50, and 75 were randomly assigned to respondents. Data analysis suggests that most visitors were willing to accept relatively high levels of risk that they would not obtain a wilderness-use permit in order to protect the park from what were judged to be unacceptable trail impacts. This suggests that trail impact would be a potentially good indicator variable, because it is a measure of the condition of park resources and is important to visitors.

CHAPTER 16

Indicators and Standards for Cultural
Resources at Mesa Verde National Park

But what does freedom mean? Freedom is the recognition of necessity.

Cultural resources are an important part of the U.S. national park system. In fact, most of the nearly four hundred units of the national park system were established to celebrate and protect important elements of the nation's history and culture. Mesa Verde National Park is an excellent example. The park was established by Congress in 1906, the same year the Antiquities Act was passed protecting historic and prehistoric buildings and objects on all federal lands. The park protects settlement areas in southwestern Colorado occupied by highly developed prehistoric cultures between AD 600 and AD 1300. The park includes over four thousand known archeological sites, including approximately six hundred cliff dwellings, for which the park is best known. The park comprises just over 50,000 acres and accommodates nearly a half million visits annually. Because of the fragile character and popularity of the cliff dwellings, they may be visited only when accompanied by a park ranger, and hiking in the backcountry of the park is not allowed.

Crowding in the Park

The popularity of the park has created concern over the issue of carrying capacity. Demand for ranger-guided tours of cliff dwellings sometimes exceeds the size and number of tours offered, and cliff dwellings that are open to self-guided tours (under the presence and supervision of park rangers) are often very heavily visited.

Initial research at the park confirmed that crowding-related issues were of concern to park visitors (Valliere and Manning 2003). Visitor questionnaires were administered at seven key sites in the park along with a survey of a representative sample of all visitors conducted at the park's only entrance/exit point. Results from a series of open- and closed-ended questions designed to identify potential indicators of the quality of the visitor experience consistently found that the most important problems for visitors were (1) getting a ticket for a ranger-guided tour, (2) too many visitors on ranger-guided tours, and (3) too many visitors in the cliff dwellings.

Crowding-Related Standards

Data on crowding-related standards were gathered at seven sites within the park. Studies at three of these sites—Cliff Palace, Spruce Tree House, and Balcony House—will be used to illustrate study methods and findings.

Cliff Palace is the largest cliff dwelling in North America and includes over one hundred rooms. For most visitors it is the icon feature of the park. This site can be visited only through participation in a one-hour, ranger-guided tour. Tickets for these tours must be purchased at the park visitor center, and demand for tickets can sometimes exceed supply.

Crowding can be manifested at Cliff Palace in two ways: the size of tour groups and the number of tour groups at the site at any one time. Visitor-based normative standards for both of these indicators were measured through a survey administered to a representative sample of visitors after they had completed their visits. Because use levels are high, a visual research approach (as described in chapter 6) was used. A series of twelve computer-edited photographs was prepared illustrating a range of one to three tour groups at the site and a range of 50 to 80 visitors in each tour. The photographs, shown in figure 16.1, were taken from the perspective of the Cliff Palace overlook where visitors gather to begin their tour. Respondents were asked to rate the acceptability of each photograph based on the number of visitors/tour groups shown.

Study findings are shown in figure 16.2. The figure includes three social norm curves—one for each series of photographs portraying different numbers of tour groups at the site. These data suggest that the number of tour groups at the site at any one time is far more important to visitors than tour-group size. (Apparently, visitors feel that multiple tour groups at the site can be disturbing.) Respondents report that one tour group at the site is very acceptable, two tour groups, marginally acceptable, and three tour groups, very unacceptable. Tour-group size is relatively unimportant, at least within the range tested. While smaller tour groups are more acceptable than larger tour groups, differences in

(a) 1 group of 50 people

(b) 1 group of 60 people

(c) 1 group of 70 people

(d) 1 group of 80 people

(e) 2 groups of 50 people

(f) 2 groups of 60 people

Figure 16.1. Study photographs illustrating a range of visitor use levels at Cliff Palace.

(g) 2 groups of 70 people

(h) 2 groups of 80 people

(i) 3 groups of 50 people

(j) 3 groups of 60 people

(k) 3 groups of 70 people

(l) 3 groups of 80 people

Figure 16.1. *Continued*

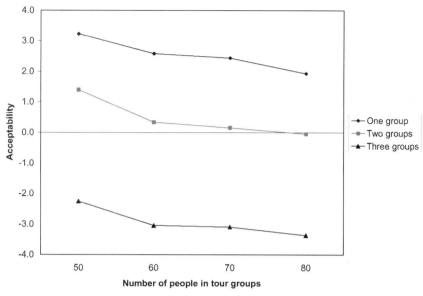

Figure 16.2. Social norm curves for number of
tour groups and tour group size at Cliff Palace.

acceptability ratings are quite small, particularly as tour-group size exceeds 60 visitors.

Spruce Tree House is also a very large cliff dwelling and is especially popular because it does not require a ticket for a ranger-guided tour. Park rangers are stationed at the site, and visitors may tour the site on their own. The site is accessible by an approximately one-quarter mile trail.

Crowding at Spruce Tree House can be manifested in two ways: at the site and along the trail leading to and from the site. Visitor-based normative standards for both of these indicators were measured through a survey administered to a representative sample of visitors as they had completed their visits. Because use levels are high, a visual research approach was used. Two series of six computer-edited photographs were prepared illustrating a range of use levels at the site and on the trail. Study photographs are shown in figures 16.3 (at the site) and 16.4 (on the trail). Respondents were asked to rate the acceptability of each photograph based on the number of visitors shown.

The social norm curves derived from study data are shown in figures 16.5 (at the site) and 16.6 (along the trail). At the site, aggregate ratings of acceptability fall out of the acceptable range and into the unacceptable range at about 53 people-at-one-time (PAOT). Along the trail, aggregate ratings of acceptability fall out of the acceptable range and into the unacceptable range at about 27 people-per-viewscape (PPV). Respondents were also asked to judge the study photo-

(a) 0 people

(b) 22 people

(c) 44 people

(d) 66 people

(e) 88 people

(f) 110 people

Figure 16.3. Study photographs illustrating a range of visitor use levels at Spruce Tree House.

graphs using several other evaluative dimensions (as discussed in chapter 5) including "preference," "management action," and "displacement," and to indicate the photograph that looked most like the levels of use that were experienced by respondents at the site and along the trail. Respondents reported that they saw an average of 42 PAOT at the site (compared to an acceptability-based threshold of 53 PAOT), and this suggests that the site is currently used at approximately

(a) 0 people

(b) 10 people

(c) 20 people

(d) 30 people

(e) 40 people

(f) 50 people

Figure 16.4. Study photographs illustrating a range of
visitor use levels on the trail to Spruce Tree House.

79% of social carrying capacity. Respondents reported that they saw an average of 20 PPV along the trail (compared to an acceptability-based threshold of 31 PPV), and this suggests that the trail is currently used at approximately 64% of social carrying capacity. Consistent with these findings, respondents reported feeling more crowded at the site than along the trail.

Based on the crowding-related standards found in this study, a computer-

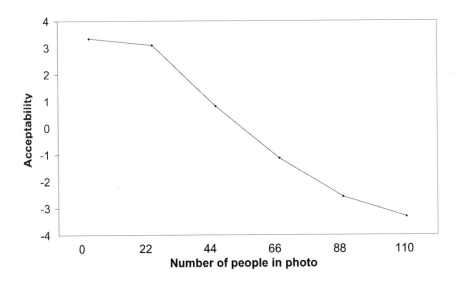

Figure 16.5. Social norm curve for PAOT at Spruce Tree House.

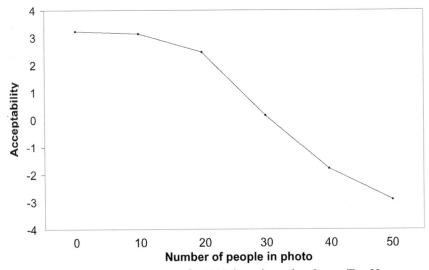

Figure 16.6. Social norm curve for PPV along the trail to Spruce Tree House.

based simulation model of visitor use (as described in chapter 8) was developed to estimate the maximum number of visitors that can be accommodated at the site. Input data for the model included daily counts of the number of visitors using the site and measures of the length of time it took visitors to hike the trail and tour the site. The model was designed to estimate the maximum number of visitors to the site and on the trail without violating acceptability-based crowding standards. The model estimates that the number of visitors per day to the site, and especially along the trail, could increase modestly to substantially without violating acceptability-based crowding standards more than 10% of the time.

Balcony House is also a large cliff dwelling, but the ranger-guided tour of this site is the most challenging in the park. The tour requires visitors to descend a hundred-foot-long staircase into the canyon, climb a thirty-two-foot ladder, crawl through a twelve-foot-long tunnel, and then climb an additional sixty feet on ladders and stone steps. A ticket is required for the tour, which lasts an hour.

Visitor-based normative standards were measured for three aspects of the ranger-guided tour: tour frequency, tour-group size, and length of tour. A visitor survey addressing these issues was administered to a representative sample of visitors as they completed their ranger-guided tour of Balcony House. The social norm curves derived from this survey are shown in figures 16.7 through 16.9.

Respondents reported that tours conducted every thirty minutes were most acceptable, that tours conducted every forty-five minutes and every sixty minutes

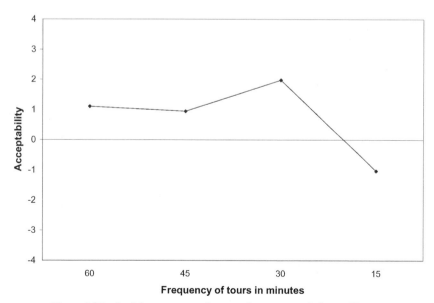

Figure 16.7. Social norm curve for tour frequency at Balcony House.

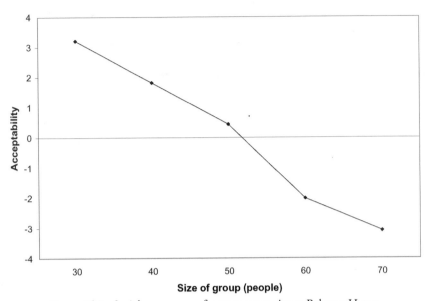

Figure 16.8. Social norm curve for tour group size at Balcony House.

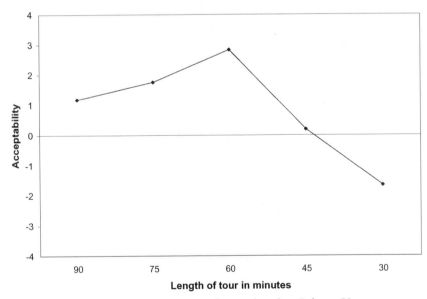

Figure 16.9. Social norm curve for tour length at Balcony House.

were also acceptable (though less so), and that tours conducted every fifteen minutes were unacceptable (figure 16.7). These findings suggest that visitors want tours conducted relatively frequently to enhance their chances of obtaining a tour ticket but do not want tours conducted so frequently that they interfere with one another.

Respondents also reported that smaller tour-group sizes are more acceptable than larger tour-group sizes (figure 16.8). Aggregate ratings fall out of the acceptable range and into the unacceptable range at about fifty-two visitors.

Finally, respondents reported that tours of sixty minutes duration were the most acceptable of the range of five tour lengths evaluated (figure 16. 9). Tours lasting seventy-five minutes and ninety minutes were also acceptable, though less so. Tours of forty-five minutes were only marginally acceptable, and tours of thirty minutes were unacceptable.

PART IV

Managing Carrying Capacity

Management is a vital component of the contemporary approach to addressing carrying capacity, as described in part 1 of this book. This approach is embodied in planning/management frameworks such as Visitor Experience and Resource Protection (VERP) and Limits of Acceptable Change (LAC). Management is needed to ensure that standards for indicator variables are maintained. Moreover, several of the research methods and subsequent case studies described in parts 2 and 3, respectively, illustrate that selected management practices can be effective in managing carrying capacity.

But what is the array of management practices that can be applied? Intuitively, limitations on the amount and type of visitor use would seem to be called for. But such limitations, by their very design and nature, have the effect of limiting public access to (and perhaps, therefore, support for) parks and protected areas. Moreover, limiting use undermines one of the primary mandates for national parks and related areas "to provide for public enjoyment." And, when recreation-related impacts to park resources and the quality of the visitor experience are caused by visitor behavior as much as by visitor numbers, limitations on use may not be fully effective in maintaining standards. Fortunately, there are other management practices that might be applied, including information and education, rules and regulations, law enforcement, zoning, and site design and management. Chapter 17 outlines these management practices, including their strategic and tactical objectives.

How effective is this array of management practices? An emerging body of research has begun to answer this question within the context of adaptive management: applying selected management practices and studying their effect. Much of this work has focused on information/education and use rationing and allocation, and study findings are summarized and presented in chapter 18.

CHAPTER 17

Alternative Management Practices

What shall we do? We have several options.

The literature on management of parks and outdoor recreation has identified a range of management practices that might be applied to issues such as crowding, conflict, environmental degradation, and other visitor-caused impacts that contribute to the issue of carrying capacity. It is useful to organize these practices into classification systems to illustrate the broad spectrum of alternatives available.

Management Strategies

One classification system defines alternatives on the basis of management strategies (Manning 1979; Manning 1999). Management strategies are basic conceptual approaches to management that relate to achievement of desirable objectives. Four basic strategies can be identified for managing parks and outdoor recreation, as illustrated in figure 17.1. Two strategies deal with supply and demand: the supply of park and recreation opportunities may be increased to accommodate more use, or the demand for recreation may be limited through restrictions or other approaches. The other two basic strategies treat supply and demand as fixed and focus on modifying either the character of recreational use to reduce its adverse impacts or the resource base to increase its durability.

There are a number of substrategies within each of these basic management strategies. The supply of parks and protected areas, for example, can be increased in terms of both space and time. With respect to space, new areas may be added or existing areas might be used more effectively through additional access or facilities. With respect to time, some park use might be shifted to off-peak periods.

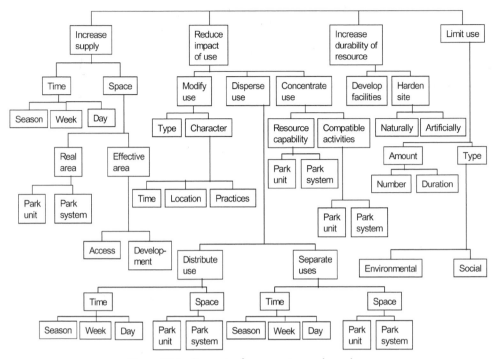

Figure 17.1. Strategies for managing parks and
outdoor recreation (from Manning, 1979).

Within the strategy of limiting demand, restrictions might be placed on the
total number of visitors that are allowed or on their length of stay. Alternatively,
certain types of use that can be demonstrated to have high environmental and/or
social impacts might be restricted.

The third basic management strategy suggests reducing the social or environmental impacts of existing use. This might be accomplished by modifying the
type or character of use, or by dispersing or concentrating use according to user
compatibility or resource capability.

A final basic management strategy involves increasing the durability of park
resources. This might be accomplished by hardening the resource itself (through
intensive maintenance, for example) or development of facilities to accommodate use more directly (wooden tent platforms, for example).

Management Tactics

A second system of classifying management alternatives focuses on tactics or on-
the-ground management practices. Management practices are direct actions or
tools applied to accomplish management strategies. Restrictions on length of

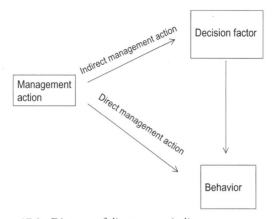

Figure 17.2. Diagram of direct versus indirect management tactics
(adapted from Peterson and Lime, 1979).

stay, differential fees, and use permits, for example, are management practices designed to accomplish the strategy of limiting recreation demand. Management practices are often classified according to the directness with which they act on visitor behavior (Gilbert et al. 1972; Lime 1977a; Peterson and Lime 1979; Chavez 1996). As the term suggests, direct management practices act directly on visitor behavior, leaving little or no freedom of choice. Indirect management practices attempt to influence the decision factors upon which visitors base their behavior. A conceptual diagram illustrating direct and indirect park and recreation management practices is shown in figure 17.2. As an example, a direct management practice aimed at reducing campfires in a wilderness environment would be a regulation barring campfires and enforcement of this regulation. An indirect management practice would be an education program designed to inform visitors of the undesirable ecological and aesthetic impacts of campfires and to encourage them to carry and use portable stoves instead. A series of direct and indirect park management practices is shown in table 17.1.

The relative advantages and disadvantages of direct and indirect park and recreation management practices have received substantial attention in the literature. Generally, indirect management practices are favored when and where they are believed to be effective (Peterson and Lime 1979; McCool and Christensen 1996). This is particularly true for wilderness and related types of park and outdoor recreation opportunities (Clark and Stankey 1979; Lucas 1982; Hendee and Dawson 2002). Indirect management practices are favored for several reasons (McCool and Christensen 1996). First, legislation and management agency policies applied to wilderness and related areas often emphasize provision of visitor opportunities that are "unconfined." Thus, direct regulation of visitor behavior may be inconsistent with such management objectives. Second, recreation is a form of leisure activity connoting freedom of choice in thought and

TABLE 17.1. Direct and indirect management practices

Type	Example
Direct (Emphasis on regulation of behavior; individual choice restricted; high degree of control)	Impose fines
	Increase surveillance of area
	Zone incompatible uses spatially (hiker-only zones, prohibit motor use, etc.)
	Limit camping in some campsites to one night or some other limit
	Rotate use (open or close roads, access points, trails, campsites, etc.)
	Require reservations
	Assign campsites and/or travel routes to each camper group in backcountry
	Limit usage via access point
	Limit size of groups, number of horses, vehicles, etc.
	Limit camping to designated campsites only
	Limit length of stay in area (maximum/minimum)
	Restrict building of campfires
	Restrict fishing or hunting
Indirect (Emphasis on influencing or modifying behavior; individual retains freedom to choose; control less complete, more variation in use possible.)	Improve (or not) access roads, trails
	Improve (or not) campsites and other concentrated use areas
	Improve (or not) fish and wildlife populations (stock, allow to die out, etc.)
	Advertise specific attributes of the area
	Identify the range of recreation opportunities in surrounding area
	Educate users to basic concepts of ecology
	Advertise underused areas and general patterns of use
	Charge consistent entrance fee
	Charge differential fees by trail, zone, season, etc.
	Require proof of ecological knowledge and recreational activity skills

Source: Adapted from Lime 1977b; Lime 1979.

actions. Regulations designed to control visitor behavior can be seen as antithet-ical to the very nature of recreation. Especially in the context of wilderness and related areas, recreation and visitor regulation have been described as "inherently contradictory" (Lucas 1982). Third, many studies indicate that, given the choice, visitors prefer indirect over direct management practices (Lucas 1983). Finally, indirect management practices may be more efficient because they do not entail the costs associated with enforcement of rules and regulations.

Emphasis on indirect management practices, however, has not been uniformly endorsed (McAvoy and Dustin 1983; Cole 1993; Shindler and Shelby 1993). It has been argued that indirect practices may be ineffective. There will always be some visitors, for example, who will ignore management efforts to influence the decision factors that lead to behavior. The action of a few may, therefore, hamper attainment of management objectives. It has been argued, in fact, that a direct, regulatory approach to management can ultimately lead to more freedom rather than less, and this is in keeping with the underlying principles of the tragedy of the commons, as described in chapter 1 (Dustin and McAvoy 1984). When all visitors are required to conform to mutually agreed-upon behavior, management objectives are more likely to be attained and a diversity of park and recreation opportunities preserved. There is empirical evidence to suggest that, under certain circumstances, direct management practices can enhance the quality of the visitor experience (Frost and McCool 1988; Swearingen and Johnson 1995). Moreover, research suggests that visitors are surprisingly supportive of direct management practices when they are needed to control the impacts of recreation use (D. Anderson and Manfredo 1986; Shindler and Shelby 1993).

An analysis of management problems caused by visitors suggests that both direct and indirect management practices can be applicable depending upon the context (Gramann and Vander Stoep 1987; Alder 1996). There are several basic reasons why visitors may not conform to desired standards of behavior. These reasons range from lack of knowledge about appropriate behavior to willful rule violations. Indirect management practices, such as information and education programs, seem most appropriate in the case of the former, while direct management practices, such as enforcement of rules and regulations, may be needed in the case of the latter.

It has been suggested that there is really a continuum of management practices that range from indirect to direct (Hendricks et al. 1993; McCool and Christensen 1996). As an example, an educational program on the ecological and aesthetic impacts of campfires would be found toward the indirect end of a continuum of management practices. A regulation requiring campers to use portable stoves instead of campfires would be a more direct management prac-

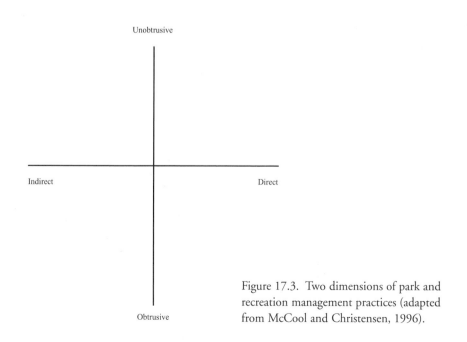

Figure 17.3. Two dimensions of park and recreation management practices (adapted from McCool and Christensen, 1996).

		Recreation management problems											
Recreation management strategies and tactics	A	B	C	D	E	F	G	H	I	J	K	L	M
I. Reduce use of the entire area													
1. Limit number of visitors in the entire area													
2. Limit length of stay in the entire area.													
3. Encourage use of other areas													
4. Require certain skills and/or equipment													
5. Charge a flat visitor fee													
6. Make access more difficult throughout the entire area													
II. Reduce use of problem areas													
7. Inform potential visitors of the disadvantages of problem areas and/or advantages of alternative areas						X		X	X	X	X		
8. Discourage or prohibit use of problem areas						X		X	X	X	X		
9. Limit number of visitors in problem areas						X		X	X	X	X		
10. Encourage or require a length-of- stay permit in problem areas						X		X	X	X	X		
11. Make access to problem areas more difficult and/or improve access to alternative areas.						X		X	X	X	X		
12. Eliminate facilities or attractions in problem areas and/or improve facilities or attractions in alternative areas.						X		X	X	X	X		
13. Encourage off-trail travel.						X			X				
14. Establish differential skill and/or equipment requirements						X		X	X	X	X		
15. Charge differential visitor fees.						X		X	X	X	X		
III. Modify the location of use within problem areas.													
16. Discourage or prohibit camping and/or stock use on certain campsites and/or locations	X		X	X	X	X	X	X	X	X	X	X	X
17. Encourage or permit camping and/ or stock use only on certain campsites and/or locations	X		X	X	X	X	X	X	X	X	X	X	X
18. Locate facilities on durable sites	X	X	X					X	X	X	X	X	X
19. Concentrate use on sites through facility design and/or information		X							X				
20. Discourage or prohibit off-trail travel		X								X	X	X	X
21. Segregate different types of visitors							X						
IV. Modify the timing of use													
22. Encourage use outside of peak use periods						X							

200

#	A	B	C	D	E	F	G	H	I	J	K	L	M
V. Modify type of use and visitor behavior.													
23. Discourage or prohibit use when impact potential is high	X	X					X	X		X	X	X	X
24. Charge fees during periods of high use and/or high-impact potential	X					X	X	X		X	X	X	X
25. Discourage or prohibit particularly damaging practices and/or equipment		X	X	X	X	X	X	X		X	X	X	X
26. Encourage or require certain behavior, skills, and/or equipment		X	X	X	X	X		X		X	X	X	X
27. Teach a recreation use ethic		X	X	X	X	X	X	X		X	X	X	X
28. Encourage or require a party size and/or stock limit			X				X	X			X	X	
29. Discourage or prohibit stock	X		X			X	X	X			X	X	
30. Discourage or prohibit pets							X			X	X	X	
31. Discourage or prohibit overnight use		X				X	X	X		X	X	X	
VI. Modify visitor expectations													
32. Inform visitors about recreation uses						X							
33. Inform visitors about conditions they may encounter in the area						X	X						
VII. Increase the resistance of the resource													
34. Shield the site from impact	X											X	
35. Strengthen the site	X							X				X	
VIII. Maintain or rehabilitate the resource													
36. Remove problems					X			X					
37. Maintain or rehabilitate impacted locations	X	X	X	X									

A. Deterioration of managed trails; B. Development of undesired trails; C. Excessive deterioration of campsites; D. Proliferation of campsites; E. Litter; F. Too many encounters; G. Visitor conflict; H. Deterioration of grazing areas; I. Human waste; J. Harassment of wildlife; K. Competition with wildlife; L. Attraction and feeding of wildlife; M. Contamination of water bodies.

Figure 17.4. Matrix of recreation management strategies and tactics and their application to recreation management problems (adapted from Cole et al. 1987).

tice. Finally, aggressive enforcement of this regulation with uniformed rangers would clearly be a very direct management practice. This suggests that management practices might also be viewed as ranging along two dimensions, as illustrated in figure 17.3 (McCool and Christensen 1996). Not only can management practices be direct or indirect, they can also be implemented in an obtrusive or unobtrusive manner. It has also been suggested that direct and indirect management practices are not mutually exclusive and that, in fact, they can often complement each other (Alder 1996; Cole et al. 1997a). For example, a regulation banning campfires (a direct management practice) should be implemented in conjunction with an educational program explaining the need for such a regulation (an indirect management practice).

A relatively comprehensive outline of park and recreation management practices is shown in the matrix in figure 17.4 (Cole et al. 1987). The vertical axis of the matrix outlines park and recreation management practices based on both strategies and tactics. Eight basic strategies are identified, and several tactics are included under each strategy. The horizontal axis outlines a series of basic problems or issues in parks and outdoor recreation. Cells within the matrix indicate the management practices that are most applicable to each type of problem or issue.

Classification of management practices might be based on many factors or concepts. The approaches described above simply illustrate the array of alternatives available for park and outdoor recreation management. For any given problem, there are likely several potential solutions. Explicit consideration should be given to this variety of approaches rather than relying on those that are simply familiar or administratively expedient.

CHAPTER 18

Evaluating the Effectiveness
of Management Practices

We must choose . . . or acquiesce in the destruction
of the commons that we call our National Parks.

Given the vital role of management in contemporary carrying capacity frameworks, it is important to test the potential effectiveness of alternative management practices. This is especially important given the array of management alternatives outlined in chapter 17. A growing body of literature has focused on evaluation of selected park and recreation management practices. This literature can be organized into several broad categories of management practices, including visitor information and education programs, use rationing and allocation, and other park and recreation management practices.

Information and Education

Information/education is generally seen as an "indirect" and "light-handed" park and recreation management tool. As noted in chapter 17, it is designed to persuade visitors to adopt behaviors that are compatible with park and recreation management objectives without regulating visitors directly, and this approach tends to be viewed favorably by park visitors (Roggenbuck and Ham 1986; Stankey and Schreyer 1987; McCool and Lime 1989; Roggenbuck 1992; Vander Stoep and Roggenbuck 1996; Hendee and Dawson 2002). Research suggests that information/education can be effective, and a set of principles for application to park and recreation management is emerging.

Conceptual and Theoretical Foundations

Problem behaviors of park visitors can be classified into five basic types, as shown in table 18.1, and this conceptual approach suggests the potential effectiveness of information/education on each. At the two ends of the spectrum, problem behaviors can be seen as either deliberately illegal or unavoidable. In these instances, information/education may have limited effectiveness. However, the other three types of problem behaviors—careless actions, unskilled actions, and uninformed actions—may be considerably more amenable to information/education programs.

Another approach to describe the application of information/education relates to the "mindfulness" or "mindlessness" of visitors (Moscardo 1999). Mindlessness relies on existing behavioral routines, and this may limit a visitor's ability to recognize and process new information. Alternatively, a mindful visitor actively processes new information, creates new categories for information, and consciously thinks about appropriate ways to behave. Strategies to enhance mindfulness can facilitate learning and better decision making (Moscardo 1999).

A third conceptual approach to the application of information/education is

TABLE 18.1. Application of information/education to management problems

Type of problem	Example	Potential effectiveness of information/education
Illegal actions	Theft of Indian artifacts; use of wilderness by motorized off-road vehicles	Low
Careless actions	Littering; shouting	Moderate
Unskilled actions	Selecting improper campsite; building improper campfire	High
Uninformed actions	Using dead snags for firewood; camping within sight or sound of another group	Very high
Unavoidable actions	Disposing of human waste; trampling groundcover vegetation at campsite	Low

Source: Adapted from Roggenbuck 1992; Vander Stoep and Roggenbuck 1996.

based on two prominent theories of moral development (Kohlberg 1976; Gilligan 1982). Both theories suggest that people tend to progress through stages of moral development, ranging from being very self-centered to highly altruistic, based on principles of justice, fairness, and self-respect. A park visitor may be at any of the stages of moral development. Management implications are that information/education should be designed to reach visitors at each of these stages (Christensen and Dustin 1989; Duncan and Martin 2002). For example, to reach visitors at lower levels of moral development, managers might emphasize extrinsic rewards or punishments for selected types of behavior. However, communicating with visitors at higher levels of moral development might be more effective by emphasizing the rationale for selected behaviors and appealing to a sense of altruism, justice, and fairness.

Fourth, communication theory suggests that the potential effectiveness of information/education is dependent upon a number of variables associated with the content and delivery of messages to visitors (Roggenbuck and Ham 1986; Stankey and Schreyer 1987; Manfredo 1989; Vaske et al. 1990; Manfredo and Bright 1991; Manfredo 1992; Roggenbuck 1992; Bright et al. 1993; Bright and Manfredo 1995; Basman et al. 1996; Vander Stoep and Roggenbuck 1996). For example, visitor behavior is at least partially driven by attitudes, beliefs, and normative standards. Information/education aimed at "connecting" with or modifying relevant attitudes, beliefs, or norms may be successful in guiding or changing visitor behavior. Moreover, the substance of messages and the media by which they are delivered may also influence the effectiveness of information/education programs.

Finally, from a theoretical standpoint, information/education can be seen to operate through three basic models (Roggenbuck 1992). *Applied behavior analysis* focuses directly on visitor behavior rather than antecedent variables such as attitudes, beliefs, and norms. For example, visitors can be informed of rewards or punishments that will be administered dependent upon their behavior. Applied behavior analysis is the simplest and most direct theoretical model of information/education. However, since it does not address underlying cognitive or behavioral variables, such as attitudes, beliefs, and norms, its effectiveness may be short term and dependent upon continued application. In the *central route to persuasion*, relevant beliefs of visitors are modified through delivery of substantive messages. New or modified beliefs then lead to desired changes in behavior. While this is a less direct and more complex model, it may result in more lasting behavioral modification. The *peripheral route to persuasion* model emphasizes nonsubstantive elements of information/education messages, such as message source and medium. For example, messages from sources considered by visitors to be authoritative or powerful may influence behavior, while other messages may be ignored. This model may be especially useful in situations where it is difficult to attract and maintain the attention of visitors, such as at visitor centers, entrance/ranger stations, and bulletin boards, all of which may offer multiple and competing information/education messages. However, like applied behavior analysis, the peripheral route to persuasion may not influence antecedent conditions of behavior and therefore may not have lasting effects.

Empirical Evaluations of Effectiveness

Empirical studies have examined the effectiveness of a variety of park and recreation-related information/education programs. These can be described as (1) studies designed to influence visitor-use patterns; (2) studies focused on enhancing visitor knowledge, especially knowledge related to minimizing ecological and social impacts; (3) studies aimed at influencing visitor attitudes toward management policies; and (4) studies that address depreciative behavior, such as littering and vandalism.

Visitor-use patterns in parks and related areas are often of uneven spatial and temporal distribution. Visitor-caused resource and experiential impacts may be reduced if use patterns could be changed. An early study in the Boundary Waters Canoe Area, Minnesota, explored the effectiveness of providing visitors with information on current use patterns as a way to alter future use patterns (Lime and Lucas 1977). Visitors who had permits for the most heavily used entry points were mailed an information packet, including a description of use patterns, noting in particular heavily used areas and times. A survey of a sample of this group who again visited the study area the following year found that three-fourths of respondents felt that this information was useful, and about one-third were influenced in their choice of entry point, route, or time of subsequent visits.

A study in the Shining Rock Wilderness Area, North Carolina, was designed to disperse camping away from a heavily used meadow (Roggenbuck and Berrier 1981; Roggenbuck and Berrier 1982). In one treatment, a brochure explained resource impacts associated with concentrated camping and showed the location of other nearby camping areas. Another group was given the brochure in addition to personal contact with a wilderness ranger. Both groups dispersed their camping activity to a greater degree than a control group, but there was no statistically significant difference between the two treatment groups.

Prior to obtaining a backcountry permit, a sample group of hikers in Yellowstone National Park (Montana, Wyoming, and Idaho) was given a guidebook that described the attributes of lesser-used trails (Krumpe and Brown 1982). Through a later survey and examination of permits, it was found that 37% of this group had selected one of the lesser-used trails compared to 14% of a control group. Results also indicated that the earlier the information was received, the more influence it had on behavior. Studies employing user-friendly, micro-computer-based information approaches (e.g., touch-screen programs) have also been found to be effective in influencing recreation-use patterns (Huffman and Williams 1987; Hultsman 1988; D. Harmon 1992; Alpert and Herrington 1998).

Hikers in the Pemigewasset Wilderness, New Hampshire, were studied to determine the influence of wilderness rangers as a source of information/education (C. Brown et al. 1992). Only about 20% of visitors reported that the information received from wilderness rangers influenced their destination within the study area. However, visitors who were less experienced and who reported that they were more likely to return to the study area were more likely to be influenced by the information provided.

Potential problems in using information/education to influence visitor use were illustrated in a study in the Selway-Bitterroot Wilderness, Montana (Lucas 1981). Brochures describing current recreation-use patterns were distributed to visitors. Follow-up measurements indicated little effect on subsequent use patterns. Evaluation of this program suggested three limitations on its potential effectiveness: (1) many visitors did not receive the brochure, (2) most of those who did receive the brochure received it too late to affect their decision making, and (3) some visitors doubted the accuracy of the information contained in the brochure.

A second category of studies has focused primarily on enhancing visitor knowledge to reduce ecological and social impacts. In Rocky Mountain National Park, Colorado, information was provided on low-impact camping practices through a series of media (Fazio 1979b). Exposure to a slide/sound exhibit, a slide/sound exhibit plus a brochure, and a slide/sound exhibit plus a trailhead sign resulted in significant increases in visitor knowledge. Exposure to a trailhead sign and brochure was not found to be very effective.

In a more recent study, a sample of day hikers to subalpine meadows in Mt. Rainier National Park, Washington, was given a short, personal interpretive program on reasons for, and importance of, complying with guidelines for off-trail hiking (Kernan and Drogin 1995). Visitors who received this program and those who did not were later observed as they hiked. Most visitors (64%) who did not receive the interpretive program did not comply with off-trail hiking guidelines, while the majority of visitors (58%) who did receive the interpretive program complied with the guidelines.

A study of day hikers at Grand Canyon National Park, Arizona, found that an aggressive information/education campaign featuring the message "heat kills, hike smart" presented in the park newspaper and on trailhead posters, influenced the safety-related hiking practices (e.g., carrying sufficient water, starting hikes early in the day) of a majority of visitors (Stewart et al. 2000). Bulletin boards at trailheads have also been found to be effective in enhancing visitor knowledge

(Cole et al. 1997a). Visitors exposed to low-impact messages at a wilderness trail-head bulletin board were found to be more knowledgeable about such practices than visitors who were not. However, increasing the number of messages posted beyond two did not result in increased knowledge levels.

Workshops and special programs delivered to organizations also can be effective in enhancing knowledge levels as well as intentions to follow recommended low-impact practices. For example, Leave No Trace (LNT) is a public/private national educational initiative that integrates outdoor recreation research into park and outdoor recreation education. LNT establishes a collaborative framework connecting managers and researchers and providing visitors with current minimum-impact skills and information (Monz et al. 1994). The effectiveness of these types of information/education programs has been demonstrated in several studies (Dowell and McCool 1986; Jones and McAvoy 1988; Cole et al. 1997a; Confer et al. 1999). Research also suggests that commercial guides and outfitters can be trained to deliver information/education programs to clients that are effective in enhancing visitor knowledge (Seig et al. 1988; Roggenbuck et al. 1992) and that trail guide booklets also can be effective (Echelberger et al. 1978).

Not all research has found information/education programs to be as effective as indicated in these studies. A study of the effectiveness of interpretive programs at Great Smoky Mountains National Park, North Carolina and Tennessee, found mixed results (Burde et al. 1988). There was no difference in knowledge about general backcountry policies between visitors exposed to the park's interpretive services and those who were not exposed. However, the former group scored higher on knowledge of park-related hazards. A test of a special brochure on appropriate behavior relating to bears found only limited change in actual or intended behavior of visitors (Manfredo and Bright 1991). Visitors requesting information on wilderness permits for the Boundary Waters Canoe Area Wilderness, Minnesota, were mailed the special brochures. In a follow-up survey, only 18% of respondents reported that they had received any new information from the brochure, and only 7.5% reported that they had altered their actual or intended behavior.

A third category of studies has examined visitor attitudes toward a variety of park and recreation management agency policies (Robertson 1982; Olson et al. 1984; Nielson and Buchanan 1986; Cable et al. 1987; Manfredo et al. 1992; Bright et al. 1993; Ramthun 1996). These studies have found that information/education can be effective in modifying visitor attitudes so they are more supportive of park and related land-management policies. For example, visitors to Yellowstone National Park (Wyoming, Montana, and Idaho) were exposed to interpretive messages about fire ecology and the effects of controlled-burn policies (Bright et al. 1993). These messages were found to influence both beliefs about these issues and attitudes based on those beliefs.

A fourth category of studies has focused on depreciative behavior, especially littering. A number of studies have found that information/education can be effective in reducing littering behavior and even cleaning up littered areas (Burgess et al. 1971; Clark et al. 1971; Marler 1971; Clark et al. 1972a; Clark et al. 1972b; Powers et al. 1973; Lahart and Barley 1975; Muth and Clark 1978; Christensen 1981; Christensen and Clark 1983; Oliver et al. 1985; Christensen 1986; Roggenbuck and Passineau 1986; Vander Stoep and Gramann 1987; Horsley 1988; Wagstaff and Wilson 1988; Christensen et al. 1992; Taylor and Winter 1995). For example, samples of visitors to a developed campground were given three different treatments: (1) a brochure describing the costs and impacts of littering and vandalism, (2) the brochure plus personal contact with a park ranger, and (3) these two treatments plus a request for assistance in reporting depreciative behaviors to park rangers (Oliver et al. 1985). The brochure plus the personal contact was the most effective treatment; this reduced the number of groups who littered their campsite from 67% to 41% and reduced the number of groups who damaged trees at their campsite from 20% to 4%. Types of messages and related purposes found to be effective in a number of studies include incentives to visitors to assist with cleanup efforts and the use of rangers and trip leaders as role models for cleaning up litter.

Related Research

Several other types of studies, while not directly evaluating the effectiveness of information/education, also suggest its potential for park and recreation management. First, studies of visitor knowledge indicate that marked improvements are possible, which could lead to improved visitor behavior. For example, campers in the Allegheny National Forest, Pennsylvania, were tested for their knowledge of the area's rules and regulations (Ross and Moeller 1974). Only 48% of respondents answered six or more of the ten questions correctly. A similar study of visitors to the Selway-Bitterroot Wilderness Area, Idaho, tested knowledge about wilderness use and management (Fazio 1979a). Only about half of the twenty questions were answered correctly by the average respondent. However, there were significant differences among types of respondents, type of knowledge, and the accuracy of various sources of information, providing indications of where and how information/education might be channeled most effectively. Visitors to the Allegheny National Forest received an average score of 48% on a twelve-item, true-false, minimum impact quiz (Confer et al. 2000), while visitors to the Selway-Bitterroot National Forest, Montana, received an average score of 33% on a similar quiz (Cole et al. 1997b).

Second, several studies indicate that information/education programs could be substantially improved (P. Brown and Hunt 1969; Fazio 1979b; Cockrell and McLaughlin 1982; Fazio and Ratcliffe 1989). Evaluation of literature mailed in

TABLE 18.2. Use and perceived effectiveness of twenty-five
information/education practices, according to wilderness managers

Practice	% of wilderness areas using these practices	Mean perceived effectiveness rating
Brochures	74	2.5
Personnel at agency offices	70	2.7
Maps	68	2.1
Signs	67	2.3
Personnel in backcountry	65	3.8
Displays at trailheads	55	2.6
Displays at agency offices	48	2.7
Posters	48	2.3
Personnel at school programs	47	2.9
Slide shows	36	2.9
Personnel at campgrounds	35	2.9
Personnel at public meetings	34	2.8
Personnel at trailheads	29	3.3
Personnel at visitor centers	26	3.0
Videos	21	2.6
Agency periodicals	18	2.3
Displays at visitor centers	18	2.5
Guidebooks	13	2.5
Interpreters	11	3.6
Computers	11	1.9
Commercial radio	9	1.9
Commercial periodicals	8	2.4
Movies	7	2.6
Commercial television	4	2.3
Agency radio	1	2.4
Mean of personnel-based techniques		3.1
Mean of media-based techniques		2.4
Mean of all techniques		2.6

Source: Adapted from Doucette and Cole 1993.

Note: Effectiveness scale: 1 = not effective; 5 = highly effective.

response to visitor requests has identified several areas of needed improvements, including more timely response, more direct focus on management problems and issues, greater personalization, more visual appeal, and reduction of superfluous materials.

Third, a survey of wilderness managers identified the extent to which twenty-five visitor information/education practices were used (Doucette and Cole 1993). Study findings are summarized in table 18.2. Only six of these practices—brochures, personnel at agency offices, maps, signs, personnel in the backcountry, and displays at trailheads—were used in a majority of wilderness areas. Managers were also asked to rate the perceived effectiveness of information/education practices. It is clear from these data that personnel-based practices are generally considered to be more effective than media-based practices.

Finally, several studies have examined the sources of information/education used by park and outdoor recreation visitors for trip planning (Uysal et al. 1990; Schuett 1993; Confer et al. 1999). Many respondents report using information/education sources that are not directly produced by management agencies, such as outdoor clubs, professional outfitters, outdoor stores, guidebooks, newspaper and magazine articles, and travel agents. This suggests that management agency linkages with selected private and commercial organizations may be an especially effective approach to information/education.

Emerging Principles for Designing and Implementing Information/Education Programs

Despite the fact that the studies are diverse in terms of geographic area, methods, and issues addressed, a number of principles for using information/education are emerging from the scientific and professional literature (Roggenbuck and Ham 1986; P. Brown et al. 1987; Manfredo 1989; Manfredo 1992; Roggenbuck 1992; Doucette and Cole 1993; Bright 1994; Basman et al. 1996; Vander Stoep and Roggenbuck 1996; Manning 2003):

- Information/education programs may be most effective when applied to problem behaviors that are characterized by careless, unskilled, or uninformed actions.
- Information/education programs should be designed to reach visitors at multiple stages of moral development.
- Information/education programs designed to "connect" with or modify visitor attitudes, beliefs, or norms are likely to be most effective in the long term, and to require less repeated application.
- Use of multiple media to deliver messages can be more effective than use of a single medium.
- Information/education programs are generally more effective with visitors who are less experienced and who are less knowledgeable.
- Brochures, personal messages, and audiovisual programs may be more effective than signs.
- Messages may be more effective when delivered early in the visitor experience, such as during trip planning.
- Messages from sources judged highly credible may be especially effective.
- Strongly worded messages and aggressive delivery of such messages can be an

effective way of enhancing the "mindfulness" of visitors and may be warranted when applied to issues such as visitor safety and protection of critical and/or sensitive resources.

- Computer-based information systems (e.g., touch-screen education programs) can be an effective means of delivering information/education.
- Messages at trailheads and bulletin boards should probably be limited to a small number of issues, perhaps as few as two.
- Training of volunteers, outfitters, and commercial guides can be an effective and efficient means of communicating information/education.
- Nonagency media, such as newspapers, magazines, and guidebooks, can be an effective and efficient means of communicating information/education.
- Information on the impacts, costs, and consequences of problem behaviors can be an effective information/education strategy.
- Role modeling by wilderness rangers and volunteers can be an effective information/education strategy.
- Personal contact with visitors by rangers or other employees can be effective in communicating information/education.
- Messages should be targeted to specific audiences to the extent possible. Target audiences that might be especially receptive include those who request information in advance and those who are least knowledgeable.
- Messages should be targeted at issues that are least well understood or known by visitors.

Studies on information/education suggest that it can be an effective and desirable management tool. Generally, these eighteen principles are based on understanding both theoretical and empirical studies reported to date, and they recommend employing a variety of message types and media and addressing a variety of management issues and target audiences.

Use Rationing and Allocation

Substantial attention has been focused on the management strategy of limiting the amount of use that parks and protected areas receive. Use rationing is controversial and is often considered to be a management approach of "last resort," because it runs counter to the basic objective of providing public access to parks and related areas (Hendee and Lucas 1973; Hendee and Lucas 1974; Behan 1974; Behan 1976; Dustin and McAvoy 1980). However, limits on use may be needed in some places and at some times to protect the integrity of critical park resources and to maintain the quality of the recreation experience.

Rationing and Allocation Practices

Five basic management practices have been identified in the literature to ration and allocate recreation use (Stankey and Baden 1977; Fractor 1982; Shelby et al. 1989; McLean and Johnson 1997): (1) reservation systems; (2) lotteries; (3) first-come, first-served, or queuing; (4) pricing; and (5) merit. A reservation system requires potential visitors to reserve a space or permit in advance of their visit. A lottery requires visitors to request a permit in advance but allocates permits on a purely random basis. A first-come, first-served, or queuing system, requires potential visitors to "wait in line" for available permits. A pricing system requires visitors to pay a fee for a permit, which may "filter out" those who are unwilling (or perhaps unable) to pay. A merit system requires potential visitors to "earn" the right to a permit by virtue of demonstrated knowledge or skill.

Each of these management practices has potential advantages and disadvantages, which are summarized in table 18.3. For example, reservation systems may tend to favor visitors who are willing and able to plan ahead, but these systems may be difficult and costly to administer. Lotteries are often viewed as eminently "fair," but they can also be cumbersome and costly to administer. Although relatively easy to administer, first-come, first-served systems may favor visitors who have more leisure time or who live relatively close to a park or related area. Pricing is a commonly used practice to allocate scarce resources in free-market economies, but it may discriminate against potential visitors with low incomes. Merit systems are rarely used but may lessen the environmental and social impacts of use.

Several principles or guidelines have been suggested for considering and applying use-rationing and allocation practices (Stankey and Baden 1977). First, emphasis should be placed on the environmental and social impacts of recreation use rather than simply on the amount of use. Some types of recreation use may cause more impacts than others. To the extent that such impacts can be reduced, rationing use of parks and recreation areas might be avoided, or at least postponed. Second, as noted, rationing use probably should be considered a management practice of last resort. Less "heavy-handed" management practices would seem more desirable where they can be demonstrated to be effective. Third, good information is needed to implement use rationing and allocation. Managers must be certain that environmental and/or social problems dictate use rationing, and that visitors are understood well enough to predict the effects of alternative allocation systems. Fourth, combinations of use-rationing systems should be considered. Given the advantages and disadvantages of each use-allocation practice, hybrid systems may have special application. For example, half of all permits might be allocated on the basis of a reservation system and half on a first-come, first-served basis. This would

TABLE 18.3. Evaluation of five park and recreation use-rationing practices

Evaluation criteria	Reservation	Lottery	First-come, first-served	Pricing	Merit
Clientele group benefited by system	Those able and/or willing to plan ahead, i.e., persons with structured lifestyles	No one identifiable group benefited; those who examine probabilities of success at different areas have better chance	Those with low opportunity cost for their time (e.g., retired, unemployed); also favors users who live nearby	Those able or willing to pay entry costs	Those able or willing to invest time and effort to meet requirements
Clientele group adversely affected by system	Those unable and/or unwilling to plan ahead: i.e., persons with occupations that do not permit long-range planning, such as many professionals	No one identifiable group discriminated against; can discriminate against the unsuccessful applicant to whom the outcome is important	Those persons with high opportunity cost of time; also those persons who live some distance from areas; the cost of time is not recovered by anyone	Those unwilling or unable to pay entry costs	Those unable or unwilling to invest time and effort to meet requirements
Experience to date with use of system	Main type of rationing system used in both U.S. national forests and national parks	Limited; however, it is a common method for allocating big-game hunting permits	Used in U.S. national parks for many services; often used with reservation systems	Little; entrance fees sometimes charged, but not to limit use	Little; merit is used to allocate use for some specialized activities such as mountain climbing and river running
Acceptability of system to users[a]	Generally high; good acceptance in areas where used; seen as best way to ration by users in areas not currently rationed	Low	Low to moderate	Low to moderate	Not clearly known; could vary considerably, depending on level of training required to attain necessary proficiency and knowledge level

214

	Rationing by request (reservation)	Rationing by lottery	Rationing by queuing	Rationing by pricing	Rationing by merit (eligibility)
Difficulty for administrators	Moderately difficult; requires extra staffing, expanded hours; record-keeping can be substantial	Difficult to moderately difficult; allocating permits over an entire use season could be very cumbersome	Low to moderate difficulty; could require development of facilities to support visitors waiting in line	Moderate difficulty; possibly some legal questions about imposing a fee for wilderness entry	Difficult to moderately difficult; initial investments to establish licensing program could be substantial
Efficiency: extent to which system can minimize problems of suboptimization	Low to moderate; underutilization can occur because of "no shows," denying entry to others; allocation of permits has little relationship to value of the experience as judged by the applicant	Low; because permits are assigned randomly, persons who place little value on an opportunity stand as good a chance of gaining entry as those who place high value on it	Moderate; because system rations primarily through a cost of time, it requires some measure of worth by participants	Moderate to high; imposing a fee requires user to judge worth of experience against costs; uncertain as to how well use could be "fine-tuned" with price	Moderate to high; requires user to make expenditures of time and effort (and maybe dollars) to gain entry
Principal way in which use is controlled	Reducing visitor numbers; controlling of distribution of use in space and time by varying number of permits available at different trailheads or at different times	Reducing visitor numbers; controlling distribution of use in space and time by number of permits available at different places or times, thus varying probability of success	Reducing visitor numbers; controlling distribution of use in space and time by number of persons permitted to enter at different places or times	Reducing visitor numbers; controlling distribution of use in space and time by using differential prices	Some reduction in numbers as well as shifts in time and space; major reduction in per capita impact
How system affects user behavior[b]	Affects both spatial and temporal behavior	Affects both spatial and temporal behavior	Affects both spatial and temporal behavior; user must consider cost of time of waiting in line	Affects both spatial and temporal behavior; user must consider cost in dollars	Affects style of user's behavior

Source: Adapted from Stankey and Baden 1977.

[a] Based upon actual field experience as well as upon evidence reported in visitor studies (Stankey 1973).

[b] This criterion is designed to measure how the different rationing systems would directly impact the behavior of users (e.g., where they go, how they behave, etc.).

serve the needs of potential visitors who can and do plan vacations in advance, as well as those whose jobs or lifestyles do not allow for this. Fifth, use rationing should establish a linkage between the probability of obtaining a permit and the value of the recreation opportunity to potential visitors. In other words, visitors who value the opportunity highly should have a chance to "earn" a permit through pricing, advance planning, waiting time, or merit. Finally, use-rationing practices should be monitored and evaluated to assess their effectiveness and fairness. Use rationing for parks and related areas is relatively new and is likely to be controversial. Special efforts should be made to ensure that use-rationing practices accomplish their objectives.

Fairness and Equity

A critical element of use-rationing and allocation practices is fairness (Dustin and Knopf 1989). Parks and outdoor recreation areas administered by federal, state, and local agencies are public resources. Use-rationing and allocation practices must be seen as both efficient and equitable. But how are equity, fairness, and related concepts defined? Several studies have begun to develop important insights into this issue. These studies have outlined several alternative dimensions of equity and measured their support among the public.

One study identified four dimensions of an overall theory of *distributive justice* (Shelby et al. 1989). Distributive justice is defined as an ideal whereby individuals obtain what they "ought" to have, based on criteria associated with fairness. A first dimension is *equality*, which suggests that all individuals have an equal right to a benefit, such as access to parks and outdoor recreation. A second dimension is *equity*, which suggests that benefits be distributed to those who "earn" them through some investment of time, money, or effort. A third dimension is *need*, which suggests that benefits be distributed on the basis of unmet needs or competitive disadvantage. A final dimension is *efficiency*, which suggests that benefits be distributed to those who place the highest value upon them.

Insights into these dimensions of distributive justice were developed through a survey of river runners on the Snake River in Hell's Canyon, Idaho (Shelby et al. 1989). Visitors were asked to rate the five use-allocation practices—reservation; lottery; first-come, first-served; pricing; and merit—on the basis of four criteria: perceived chance of obtaining a permit, perceived fairness of the practice, acceptability of the practice, and willingness to try the practice. Results suggest that visitors use concepts of both fairness and pragmatism in evaluating use-rationing practices. However, pragmatism—the perceived ability on the part of the respondent to obtain a permit—has the strongest effect on

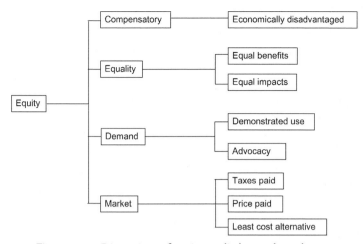

Figure 18.1. Dimensions of equity applied to parks and recreation
(adapted from Crompton and Lue 1992).

willingness to try each of the allocation practices. These findings suggest that managers have to convince potential visitors that proposed use-allocation practices are not only fair but that they will provide them with a reasonable chance to obtain access.

A second series of studies has examined a more extended taxonomy of equity dimensions that might be applied to provision of a broad spectrum of park and recreation opportunities (Wicks and Crompton 1986, 1987, 1989, 1990; Wicks 1987; Crompton and Wicks 1988; Crompton and Lue 1992). Eight potential dimensions of equity are identified, as shown in figure 18.1. A first dimension is compensatory and allocates benefits on the basis of economic disadvantage. The second two dimensions are variations of equality and allocate benefits to all individuals equally, or ensure that all individuals ultimately receive equal total benefits. The fourth and fifth dimensions are based on demand and allocate benefits to those who make greatest use of them or those who advocate most effectively for them. The final three dimensions of equity are market driven and distribute benefits based on amount of taxes paid, the price charged for services, or the least-cost alternative for providing recreation services.

These dimensions of equity were described to a sample of California residents, and respondents were asked to indicate the extent to which they agreed or disagreed with each dimension of equity as a principle for allocating public park and recreation services (Crompton and Lue 1992). A majority of the sample agreed with only three of the dimensions. These dimensions were, in decreasing order, demonstrated use, price paid, and equal benefits.

Visitor Attitudes and Preferences

Despite the complex and controversial nature of use rationing and allocation, there appears to be considerable support for a variety of such management practices among park and outdoor recreation visitors (Stankey 1973; Stankey 1979; Fazio and Gilbert 1974; Lucas 1980; Lucas 1985; McCool and Utter 1981; McCool and Utter 1982; Utter et al. 1981; Shelby et al. 1982; Shelby et al. 1989; Schomaker and Leatherberry 1983; Glass and More 1992; Watson 1993; Watson and Niccolucci 1995). Research suggests that even most individuals who have been unsuccessful at obtaining a permit continue to support the need for use rationing (Fazio and Gilbert 1974; Stankey 1979; McCool and Utter 1982). A study of visitors to three wilderness areas in Oregon found that support for use restrictions was based on concerns for protecting both resource quality and the quality of the visitor experience (Watson and Niccolucci 1995). Support by day hikers was influenced most strongly by concerns with crowding, while support by overnight visitors was influenced by concern for both crowding and environmental impacts.

Preferences among alternative use-rationing practices have been found to be highly variable, based on both location and type of user (Magill 1976; McCool and Utter 1981; Shelby et al. 1982; Shelby et al. 1989; Glass and More 1992). Support for a particular use-allocation practice appears to be related primarily to which practices respondents are familiar with, and the extent to which they believe they can obtain access. A study of river managers found that first-come, first-served and reservation systems were judged the two most administratively feasible allocation practices and were also the most commonly used practices (Wikle 1991).

In keeping with the generally favorable attitude toward use limitation, most studies have found visitor compliance rates for mandatory permits to be high, ranging from 68% to 97%, with most areas in the 90% range (Lime and Lorence 1974; Godin and Leonard 1977; van Wagtendonk and Benedict 1980; Plager and Womble 1981; Parsons et al. 1982). Moreover, permit systems that have incorporated trailhead quotas have been found to be effective in redistributing use both spatially and temporally (Hulbert and Higgins 1977; van Wagtendonk 1981; van Wagtendonk and Coho 1986).

Pricing

Among the use-rationing and allocation practices, pricing has received special attention in the literature. Pricing is the primary means of allocating scarce resources in a free market economy. Economic theory generally suggests that higher prices will result in less consumption of a given good or service. Thus, pricing may be an effective approach to limiting use of parks and related areas. However, park and recreation services in the public sector traditionally have been priced at a nominal level or have been provided free of charge. The basic philosophy underlying this policy is that access to park and recreation services is important to all people, and no one should be "priced out of the market." Interest in instituting or increasing fees at parks and outdoor recreation areas has generated a considerable body of literature that ranges from philosophical to theoretical to empirical (F. Anderson and Bonsor 1974; Gibbs 1977; Manning and Baker 1981; Driver 1984; Manning et al. 1984; Manning et al. 1996f; Rosenthal et al. 1984; Cockrell and Wellman 1985a; Cockrell and Wellman 1985b; Dustin 1986; Manning and Koenemann 1986; Martin 1986; McCarville et al. 1986; Walsh 1986; Daniels 1987; Dustin et al. 1987; Harris and Driver 1987; Leuschner et al. 1987; McCarville and Crompton 1987; McDonald et al. 1987; Wilman 1988; Bamford et al. 1988; Reiling et al. 1988; Reiling et al. 1992; Reiling et al. 1996; Schultz et al. 1988; Fedler and Miles 1989; Stevenson 1989; Manning and Zwick 1990; Kerr and Manfredo 1991; G. Peterson 1992; Christensen et al. 1993; Reiling and Cheng 1994; Scott and Munson 1994; Emmett et al. 1996; Lundgren 1996; McCarville 1996; Reiling and Kotchen 1996; Bowker and Leeworthy 1998).

Studies of pricing have tended to focus on several issues related to its potential as a recreation management practice. First, to what extent does pricing influence use of parks and related areas? Several studies have found an inverse relationship between price and use (Lindberg and Aylward 1999; Richer and Christensen 1999; Schroeder and Louviere 1999). For example, a study of day users at six recreation areas administered by the U.S. Army Corps of Engineers found that 40% of respondents reported they would no longer use these areas if a fee was instituted (Reiling et al. 1996). However, other studies have shown little or no effects of pricing on visitor-use levels (Manning and Baker 1981; Becker et al. 1985; Leuschner et al, 1987; Rechisky and Williamson 1992). The literature suggests that the influence of fees on park and recreation use is dependent upon several factors.

- The *elasticity* of demand for a park or recreation area. Elasticity refers to the slope of the demand curve that defines the relationship between price and quantity consumed (or visitation). This issue is illustrated in figure 18.2. The demand for some recreation areas is relatively elastic, meaning that a change in price has a comparatively large effect on visitation. The demand for other recreation areas is relatively inelastic, meaning that a change in price has a comparatively small effect on visitation.
- The significance of the park or recreation area. Parks of national significance, such as Yellowstone National Park, are likely to have a relatively inelastic demand, suggesting that pricing is not likely to be effective in limiting use

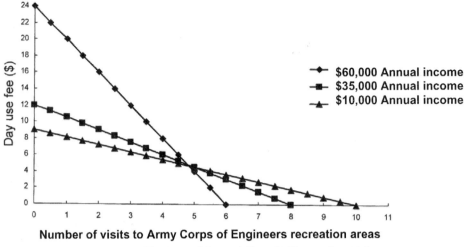

Figure 18.2 Elasticity of demand (from Reiling et al., 1996).

unless price increases are dramatic. Less significant parks are likely to be characterized by more elastic demand, and pricing may be an effective use-allocation practice.

• The percentage of total cost represented by the fee. In cases where the fee
  charged represents a relatively high percentage of the total cost of visiting a
  park or recreation area, pricing is likely to be a more effective use-limiting
  approach. However, when the fee charged represents only a small percentage
  of the total cost, pricing is not likely to be an effective use-limiting approach.

• The type of fee instituted. Pricing structure can be a potentially important element in determining the effectiveness of fees as a management practice. For
  example, a daily-use fee might be more effective in limiting total use than an
  annual pass that allows unlimited use opportunities for a flat fee.

   A second issue addressed in the literature is the acceptability of fees to potential visitors. Again, study findings are mixed, though they often suggest that there
is a substantial willingness to pay for park and recreation services (Bowker et al.
1999; Krannich et al. 1999; Vogt and Williams 1999; Williams et al. 1999; Winter et al. 1999). However, research suggests that the acceptability of fees is at least
partially dependent on several factors.

• Dispensation of resulting revenues. If revenues derived from fee programs are
  retained by the collecting agency and reinvested in recreation facilities and
  services, then fees are often judged to be more acceptable to park visitors.

• Initiation of fee or increase in existing fee. Public acceptance of new fees where none were charged before tends to be relatively low compared to increases in existing fees.

- Local or nonlocal visitors. Local visitors tend to be more resistant to new fees than nonlocal visitors. As described earlier, this is probably because fees represent a larger percentage of the total cost of visiting a recreation area for local visitors. Moreover, local residents are more likely to visit a given recreation area more often than nonlocal residents.
- Provision of comparative information. Visitor acceptance of a fee is likely to be greater when information is provided on the costs of competing or substitute recreation opportunities and when visitors are made aware of the costs of providing recreation opportunities.

A third issue concerns the potential for pricing to discriminate against certain groups in society, particularly those with low incomes and minority racial and ethnic groups. Once again, research on this issue is mixed. For example, one study examined the socioeconomic characteristics of visitors to two similar outdoor recreation areas in Virginia, one of which charged an entrance fee and one of which did not (Leuschner et al. 1987). No differences were found in income levels, suggesting the fee had no discriminatory effect. However, several studies have found evidence of discriminatory effects (Bowker and Leeworthy 1998; Bowker et al. 1999; Schneider and Budruk 1999). For example, two studies of willingness to pay recreation fees at state parks and U.S. Army Corps of Engineers day-use areas found that lower-income visitors had a more elastic demand curve than did high-income users, as illustrated in figure 18.2 (Reiling et al. 1992; Reiling and Cheng 1994). This suggests that pricing may discriminate against lower-income visitors.

A final issue concerns the use of differential pricing to influence park and recreation use patterns. Differential pricing consists of charging higher or lower fees at selected times and locations. Research demonstrates that outdoor recreation tends to be characterized by relatively extreme "peaking." That is, certain areas or times are used very heavily, while other times or areas are relatively lightly used. Can pricing be used to even out such recreation use patterns? Research is suggestive of this potential use of pricing (LaPage et al. 1975; Willis et al. 1975; Manning et al. 1982). For example, studies of experimental differential campsite pricing at state parks in Vermont documented significant shifts in campsite occupancy patterns (Manning et al. 1984; Bamford et al. 1988).

Other Park and Recreation Management Practices

As suggested in chapter 17, a number of other practices are available to manage parks and outdoor recreation. Most of these tend to be direct management practices. Beyond information/education programs and limiting use, four broad categories of management practices addressed in the literature include rules and regulations, law enforcement, zoning, and site design and management.

Rules and Regulations

Rules and regulations are a commonly used recreation management practice, though their use sometimes can be controversial (Lucas 1982; Lucas 1983). Common applications of rules and regulations in parks and outdoor recreation include group-size limitations, assigned campsites and/or travel itineraries, area closures, length of stay limitations, and restrictions on and/or prohibition of campfires. The importance of encouraging visitors to comply with rules and regulations is emphasized in a study of the national park system that found that visitors who did not comply with rules and regulations caused extensive damage (Johnson and Vande Kamp 1996).

As noted earlier in this chapter, research indicates that visitors are often unaware of rules and regulations (Ross and Moeller 1974). This suggests that managers must effectively communicate rules and regulations to visitors using the principles and guidelines described in the section on information and education programs. In particular, visitors should be informed of the reasons why applicable rules and regulations are necessary, sanctions associated with failure to comply with rules and regulations, and alternative activities and behaviors that can be substituted for those not allowed.

Only limited research has addressed the effectiveness of rules and regulations as a recreation management practice. The literature suggests most visitors support limitations on group size but that group types should also be considered when promulgating such regulations (Roggenbuck and Schreyer 1977; Heywood 1985). Group-size limits should not be set so low that they affect primary social groups of visitors (e.g., families, close friends) who may have strong motivations for social interaction.

Research suggests that regulations requiring the use of assigned campsites in wilderness or backcountry are generally not supported by visitors (Lucas 1985; D. Anderson and Manfredo 1986). A version of this regulation requires backpackers to follow a fixed travel itinerary. Studies of the effectiveness of this regulation have found that visitor compliance rates are relatively low (van Wagtendonk and Benedict 1980; Parsons et al. 1981; Parsons et al. 1982; Stewart 1989; Stewart 1991). For example, 44% to 77% of backcountry campers were found not to be in full compliance with their permit itinerary across four zones of Grand Canyon National Park, Arizona (Stewart 1989). Noncompliance was caused primarily by visitors using campsites other than those specified or staying in the backcountry more or fewer nights than originally specified.

Research on regulations closing selected areas to public use suggest they are supported by visitors if the underlying reason is clear and justified (Frost and

McCool 1988). Most visitors were found to obey a regulation closing selected backcountry campsites for ecological reasons (Cole and Rang 1983).

Law Enforcement

Little research has been conducted on law enforcement in parks and outdoor recreation. Most of the literature in this area discusses the controversial nature of law enforcement in this context (Campbell et al. 1968; Bowman 1971; Hadley 1971; Hope 1971; Schwartz 1973; Connors 1976; Shanks 1976; Wicker and Kirmeyer 1976; L. Harmon 1979; Morehead 1979; Wade 1979; Westover et al. 1980; Philley and McCool 1981; Heinrichs 1982; Perry 1983; Manning 1987). However, one study focused on the use of uniformed rangers to deter off-trail hiking at Mt. Rainier National Park, Washington (Swearingen and Johnson 1995). The presence of a uniformed ranger was found to significantly reduce off-trail hiking. Moreover, visitors tended to react positively to this management practice when they understood that the presence of a ranger was needed for information dissemination, visitor safety, and resource protection.

Zoning

Zoning is another basic category of recreation management practices. In its most generic sense, zoning simply means assigning certain recreation activities to selected areas (or restricting activities *from* areas, as the case may be). Zoning also can be applied in a temporal dimension as well as in a spatial sense. Finally, zoning can be applied to alternative management prescriptions as a way of creating different types of park and outdoor recreation opportunities (Greist 1975; Haas et al. 1987). For example, "rescue" and "no-rescue" zones have been proposed for wilderness areas, though this is controversial (McAvoy and Dustin 1981; McAvoy et al. 1985; Dustin et al. 1986; Harwell 1987; D. Peterson 1987; McAvoy 1990).

In its most fundamental form, zoning is widely used to create and manage a diversity of park and recreation opportunities. Zoning also is used in outdoor recreation to restrict selected recreation activities from environmentally sensitive areas and to separate conflicting recreation uses. No primary research has been conducted on the potential effectiveness of zoning.

Site Design and Management

A final category of park and recreation management practices is site design and management. Recreation areas can be designed and manipulated to "harden" them against recreation impacts and manage their use. For example, boardwalks

can be built to concentrate use in developed areas, and facilities can be constructed along trails to channel use in appropriate areas (Hultsman and Hultsman 1989; Doucette and Kimball 1990). Moreoever, campsites can be designated and designed in ways to minimize social and ecological impacts (Godin
and Leonard 1977b; McEwen and Tocher 1976; Echelberger et al. 1983). However, most of these management practices involve resource management and
activities that may not be appropriate in some parks and protected areas. Hammitt and Cole (1998) provide an excellent review of the outdoor recreation literature addressing environmental impacts of outdoor recreation and associated
site and resource management practices.

Status and Trends in Park and Recreation Management

What park and recreation management practices are used most often? How
effective do managers think these practices are? What are the trends in park and
recreation management? Several studies conducted over the past few decades
offer insights into these questions (Godin and Leonard 1979; Bury and Fish
1980; Fish and Bury 1981; Washburne 1981; Washburne and Cole 1983; Marion et al. 1993; Manning et al. 1996c; Abbe and Manning 2005). These studies
have focused on wilderness and backcountry areas and have involved periodic
surveys of park and protected-area managers. A recent study explored recreation
management practices in the national park system (Marion et al. 1993; Manning
et al. 1996c). Managers of all national park backcountry areas were asked to indicate which of more than one hundred recreation management practices were currently used and which were judged most effective. Management practices used
in over half of all areas are shown in table 18.4, along with all management practices judged to be "highly effective."

Comparisons across the studies can provide some insights into trends in park
and recreation management problems and practices, at least in the context of
wilderness and backcountry areas. Although the areas, management agencies,
and research methods varied among these studies, their primary objectives were
similar—to assess recreation management problems and/or practices in parks
and recreation areas. These studies provide benchmarks at five points in time—
1979, 1981, 1983, 1993, and 2004—and suggest several basic trends in park
and recreation management problems and practices.

First, environmental impacts, primarily on trails and campsites, are the dominant recreation-related problems perceived by managers throughout these studies. In all five studies, managers tended to report site deterioration, including soil
erosion and loss of vegetation, as the most frequently occurring recreation management problem.

TABLE 18.4. Most commonly used and most effective
park and recreation management practices

Most commonly used (% of areas using)	Most effective
Educate visitors about "pack-it-in, pack-it-out" policy (91)	Campsite impacts • Designate campsites
Prohibit visitors from cutting standing dead wood for fires (83)	• Prohibit campfires • Provide campsite facilities
Educate visitors about how to minimize their impacts (77)	• Restore campsites • Limit group sizes
Remove litter left by visitors (74)	• Implement campsite reservation system
Instruct visitors not to feed wildlife (74)	
Require backcountry overnight visitors to obtain permits (68)	Trail impacts • Maintain and rehabilitate trails
Instruct visitors to bury human wastes (66)	• Use impact-monitoring system
Require groups to limit their length of stay at campsites (64)	• Use formal trail system and plan • Implement quotas on amount of use
Give verbal warnings to visitors who violate regulations (63)	Wildlife impacts
Require groups to limit their size (62)	• Temporarily close sensitive areas
Prohibit pets from the backcountry (61)	• Regulate food storage and facilities
Prohibit use of horses in selected areas (59)	• Provide user educational program
Instruct visitors to bury human wastes away from all water sources (57)	• Prohibit or restrict pets • Provide information workshops for commercial outfitters and guides
Inform visitors about potential crowding they may encounter in selected areas (56)	
Discourage use of environmentally sensitive areas (54)	Water impacts • Provide primitive toilets at high-use sites
Inform visitors about managers' concerns with visitor-use impacts at attraction areas (54)	
Instruct visitors to view wildlife from a distance (53)	Visitor crowding and conflicts • Implement quotas on amount of visitor use
Perform regular trail maintenance (52)	
Require groups to limit their length of stay in the backcountry (51)	• Control access to backcountry with visitor transportation system

Source: Adapted from Manning et al. 1996c.

Second, social problems of crowding and conflicting uses appear to have increased over time. The initial study in 1979 revealed no crowding problems. The study reported that user conflict was cited as a problem by 29% of wilderness managers, but this conflict was associated primarily with nonconforming uses of wilderness, such as grazing by domestic livestock and off-road vehicles. More recent studies report substantial and increasing levels of crowding and conflict among recreation users. For example, crowding was reported as a problem "in many places" in 1983 at 10% of all areas studied, including 2% of National Park Service (NPS) areas. By 1993, between 10% and 27% (depending upon location—campsite, trail, attraction site—within the area) of NPS areas reported crowding "in many or most areas." Moreover, conflict between different types of users was reported as widespread in 2% of areas in 1983 but was reported as a problem "in many or most areas" in 1993 by as many as 9% of areas.

Third, carrying capacity has become a pervasive but largely unresolved issue. The initial study in 1979 did not report carrying capacity as a significant issue. However, by 1983, recreation use was judged to exceed carrying capacity "sometimes" or "usually" in at least some areas by more than half of all managers. Carrying capacity problems in NPS areas were reported as equally extensive in 1983 and 1993, with 70% of managers reporting that carrying capacity is exceeded either "sometimes" or "usually" in at least some areas. Despite the apparent seriousness of the carrying capacity issue, most managers have not yet addressed it adequately. Nearly half of all areas studied in 1983 reported that they were unable to estimate carrying capacity for any portions of their areas. Moreover, the percentage of NPS areas unable to estimate carrying capacity rose from 36% in 1983 to 57% in 1993. The vast majority (76.4%) of NPS wilderness managers reported in 2004 that carrying capacity for day-use visitors had not been estimated in their areas. Finally, despite the fact that 43% of NPS areas were able to estimate carrying capacity in at least some portions of their areas in 1993, considerably less than half of these areas make such estimates based on scientific studies.

Fourth, implementation of both direct and indirect recreation management practices have tended to increase over time. For example, overnight permits for backcountry camping were required by 41% of areas in 1983, but were required by 68% of areas in 1993. Party-size limits are imposed in increasing numbers of areas, up from 43% in 1981 to 62% in 1993. Length-of-stay limits are also imposed in increasing numbers of areas, up from 16% in 1981 to 51% in 1993. Finally, minimum-impact education programs were employed in 77% of areas in 1993, up from 35% reported in 1981. Although some of these differences may be the result of differences among management agencies, the magnitude suggests a shift in management practices.

Fifth, day use is an emerging issue that warrants more management attention. The study in 1983 was the first to report that a very large percentage of all wilderness-related recreation use was accounted for by day users. The average percentage of all visitor groups that are day users ranged from 44% in Bureau of Land Management areas to 83% for Fish and Wildlife Service areas. In NPS areas, the percentage of day users has remained relatively constant over the past two decades: 62% in 1983, 64% in 1993, and 57% in 2004. The issue of day use is exacerbated by two factors (Roggenbuck et al. 1994). First, many management problems are attributed by managers to day users. In fact, in the judgment of managers, day users are more responsible than overnight visitors for most types of management problems. Second, day users often are not targeted for management actions. For example, less than 10% of NPS areas require a permit for day use (Abbe and Manning 2005).

Finally, management of parks and outdoor recreation is becoming more complex and more sophisticated. This trend is reflected in the nature of the five studies examined in this section. The original study in 1979 was primarily an exploratory study asking managers to describe their important problems. Simply defining the basic concept of wilderness areas emerged as a important issue while managers struggled with the legal and operational definitions of wilderness and related areas. The second study, reported in 1981, focused primarily on recreation management practices across several land-management agencies. The third study in 1983 adopted several objectives, including determining recreation-use patterns, recreation-related problems, and recreation management practices. The fourth study incorporated the preceding objectives and added others, including investigating the perceived causes of management problems, the effectiveness of management practices, and the degree to which management actions are based on scientific study. The fifth and most recent study, in 2004, replicated much of the 1993 study to monitor trends in back-country/wilderness management and expanded this study to focus more directly on the emerging issue of day use. The progression of these five studies illustrates that awareness and knowledge about park and recreation-related problems and management practices are expanding.

Studies on alternative recreation management practices are beginning to be marshaled into handbooks and other types of guidelines that can be used by park and recreation managers. For example, the matrix illustrated in figure 17.4 in chapter 17 is adapted from a handbook developed for wilderness managers (Cole et al. 1987). In addition to suggesting which recreation management practices might be applied to a series of recreation-related problems, the handbook offers basic information on understanding and applying each of the thirty-seven recreation management practices identified. A similar handbook has been developed for use by managers of national parks and related areas (D. Anderson et al. 1998). Prototypes of computer-based "expert systems" also are being developed to provide park and recreation managers with guidance based on the scientific and professional literature (Flekke et al. 1996).

PART V

Beyond Parks and Protected Areas

The conceptual frameworks and research methods described in parts I and II of this book offer approaches to analyzing and managing the carrying capacity of parks and protected areas. Moreover, the case studies described in part 3 illustrate ways in which these approaches might be applied to an array of park contexts. Part 4 outlines a range of actions that can be used to manage the carrying capacity of parks and protected areas. But there are many other manifestations of the tragedy of the commons beyond parks and protected areas and many other broad-ranging environmental issues that need to be managed.

The conceptual, research, and management approaches outlined in this book might be applied more broadly in at least two ways. First, these approaches can be used to manage parks and protected areas in ways that extend beyond the issue of carrying capacity. Visitor Experience and Resource Protection (VERP), Limits of Acceptable Change (LAC), and related frameworks are fundamentally management-by-objectives systems: they require that management objectives be formulated in specific, measurable terms; that relevant resource and experiential conditions in parks be monitored; and that management practices be applied to help ensure that management objectives are achieved and maintained. This offers a thoughtful and systematic approach to managing many environmental and social issues in parks and related areas.

Second, this approach can be applied to broad-ranging environmental management issues and areas. Carrying capacity of parks and protected areas addresses the fundamental tension between use of these areas and protection of important resources and experiential conditions. This is a specific manifestation of the broader tension between (1) the degree to which we can use the environment for a host of purposes, and (2) protecting what we find valuable about it. The conceptual framework outlined in this book, and the research methods designed to support its application, can be useful in guiding environmental management in an array of manifestations and contexts.

In fact, this approach is now being integrated into many applications of environmental management. For example, indicators (and to a lesser degree, standards) now guide planning and management in many environmental fields. Moreover, evolving contemporary environmental management concepts—such as ecosystem management, adaptive management, and sustainability—are highly compatible with the conceptual, research, and management approaches described in this book and applied in the context of parks and protected areas. This final chapter reviews this work in several environmental fields and concludes with an international case study of the Lake Champlain Basin.

CHAPTER 19

Indicators and Standards of Sustainability

The logic of the commons has been understood for a long time. . . .
But it is understood mostly . . . in special cases which are not sufficiently generalized.

Contemporary environmental management is being guided by a number of emerging concepts, including ecosystem management, sustainability, and adaptive management. While definitions and operational procedures for these concepts are still evolving, several principles can be isolated that might be broadly applicable to many forms of environmental management. First, environmental management must address the integration of ecology and society (Agee and Johnson 1987; Society of American Foresters 1993; Grumbine 1994). The integrity of important ecological processes must be protected, but natural and environmental resources also must be managed for the benefits of society. Thus, ecosystem management has been defined as "regulating . . . ecosystem structure and function . . . to achieve socially desirable conditions" (Agee and Johnson 1987) and "integrating . . . ecological relationships within a complex sociopolitical and values framework" (Grumbine 1994).

Second, managing the environment for the benefits of the present generation should not preclude the ability of future generations to attain needed environmentally related benefits. This principle is at the heart of the emerging concept of sustainability as originally outlined by the World Commission on Environment and Development (1987) and as sustainability is now being applied in many environmental and related fields.

Third, environmental management should be conducted within a framework that identifies goals and objectives and works toward these ends through a program of monitoring and management. A recent report by the Ecological Society of America recommends that environmental management be "driven by explicit

231

goals . . . and made adaptable by monitoring and research" (Christensen et al. 1996). This principle is fundamental to the evolving concept of adaptive management, which emphasizes the role of ongoing monitoring and evaluation as a way of informing environmental management (Lee 1993; Stankey et al. 2005; Holling 1978; Walters 1986).

The carrying capacity–related frameworks and research approaches described in this book can be useful in applying the concepts and principles of contemporary environmental management. Visitor Experience and Resource Protection (VERP), Limits of Acceptable Change (LAC), and related management frameworks offer a procedural approach that emphasizes (1) the development of management objectives (often called desired conditions in the context of parks and protected areas) and their expression in the form of quantitative indicators and standards; (2) a long-term commitment to monitoring indicator variables; and (3) a program of management that responds to monitoring data and is designed to ensure that standards for indicator variables are maintained. This procedural framework addresses both resource (or ecological) and experiential (or social) aspects of environmental management. Finally, this procedural framework was developed explicitly to address the inherent tension between resource use and protection, or the underlying issue of sustainability: how much can we use the environment without diminishing its ecological and social value to an unacceptable degree? Thus, the conceptual foundations of the carrying capacity frameworks described in this book may offer a useful approach to guiding broader environmental management.

The research approaches described and illustrated in this book also can contribute to guiding this broader environmental management. Research designed to help formulate indicators and standards can be applied in an array of environmental contexts in addition to parks and protected areas. Society has a potentially important role to play in identifying indicators of environmental quality. Moreover, people may well have normative standards for many of the environmental conditions they encounter and value in their everyday lives. Visual research methods may be helpful in some of these contexts. Research on inherent tradeoffs between resource use and protection are nearly universal in their application to issues of sustainability. And innovative approaches to monitoring (such as simulation modeling) are needed to guide management actions designed to help ensure that standards are maintained and to assess the effectiveness of these management actions.

Environmental Indicators and Standards

Application of the emerging concepts and principles of environmental management is most clearly manifested in the development and use of environmental and social indicators and, to a lesser degree, standards. The contemporary scientific and professional literature contains thousands of references to the expanding use of environmental and social indicators (see, for example, National Research Council 2000; Niemi and McDonald 2004; McKenzie et al. 1992; U.S. Environmental Protection Agency 2002 for reviews of this literature). There is evidence that early humans relied on environmental indicators such as migratory animal movements for information about changing natural conditions (Niemi and McDonald 2004). However, modern scientific use of environmental indicators can be traced to the work of Clements (1920), who laid the foundation for the use of plants as indicators of ecological conditions and processes (Morrison 1986). Perhaps the most widely known early use of environmental indicators was "the canary in the coal mine" as a measure of air quality (Burrell and Siebert 1916). Environmental indicators have expanded to include a host of measures other than observation of plant and animal species, and they sometimes use indexes comprising multiple variables.

Social indicators also have a relatively long history of use. An early example is the work of H. Odum (1936), who developed a large suite of indicators of socioeconomic conditions in the southern United States for purposes of regional planning (Force and Machlis 1997). Economic indicators such as unemployment rate, interest rate, and gross national product (GNP), along with social indicators such as crime rate, literacy, and life expectancy have been central to economic and social planning in the United States for many years. Emergence of the concept of ecosystem management has emphasized the connections between the environment and society, and this has suggested that environmental management should include indicators of both ecological and associated social conditions.

Contemporary emphasis on the use of indicators is tied to the concept of sustainability and is a direct outgrowth of the United Nations Conference on Environment and Development (popularly known as the Earth Summit) held in Rio de Janeiro in 1992. This conference prepared a plan of action titled *Agenda 21* to achieve sustainability on a global basis and called for identification of "indicators of sustainable development." The Commission on Sustainable Development was established to help ensure effective follow-up. To monitor the implementation of Agenda 21, the commission established 134 (more recently reduced to 57) broad-ranging indicators (Commission on Sustainable Development 2001). The list includes environmental (e.g., ambient concentration of air pollutants in urban areas), social (e.g., population with access to safe drinking water), and institutional/managerial (e.g., implementation of national sustainable development strategy) variables.

The work of the Commission on Sustainable Development has been extended to many areas of environmental management by a host of governmental and nongovernmental organizations. For example, one of the more highly developed applications of indicator-based approaches to environmental management is the current program of sustainable forestry. In 1993, following the Earth Summit, the International Seminar of Experts on Sustainable Development of Boreal and Temperate Forests was held in Montreal. A further outgrowth of this initiative was the Working Group on Criteria and Indicators for the Conservation and Sustainable Management of Temperate and Boreal Forests, popularly known as the *Montreal Process Working Group*. In 1995, in its meeting in Santiago, the working group developed seven criteria and sixty-seven indicators (popularly known as the *Santiago Declaration*) to guide sustainable forestry at the country or national level. The seven criteria are analogous to management objectives or desired conditions as conceived in contemporary carrying capacity frameworks. For example, the first criterion is *conservation of biological diversity*. The sixty-seven indicators are measurable, manageable variables that can be used as proxies for these criteria or objectives. For example, indicators of the first criterion include the number of forest-dependent species and extent of area by forest type relative to total forest area. The seven criteria included in the Montreal Process range from ecological to social to institutional considerations. The criteria and indicators included in this program are intended to provide a commonly agreed-upon understanding of what is meant by sustainable forest management and to be a mechanism for evaluating a country's success at achieving sustainability at the national level. Given substantive differences among nations regarding basic forest-related conditions (e.g., amount of forest land, population density), standards for indicator variables are left to the discretion of countries that choose to endorse the Santiago Declaration. These countries are expected to monitor indicators on a regular basis, with resulting data suggesting the degree to which sustainability in forest management is being achieved, and informing national and international forestry-related policy and management.

The Lake Champlain Basin: A Case Study in Environmental Management

The Lake Champlain Basin provides a current example of environmental management that is being guided by the concepts and principles described in this chapter and that has employed some of the research methods described in earlier parts of this book (Watzin et al. 2005; Smyth et al., forthcoming). A series of management objectives for the ecological and related social conditions of this area has been developed through a program of research and public policy. However, indicators for these objectives must be identified and prioritized, and standards for these indicators must be formulated. This section outlines how this work is proceeding based on a program of ecological and social science.

Lake Champlain is a transboundary lake with substantial ecological, recreational, economic, and historical values. Because of its location between New York and Vermont, U.S., and Quebec, Canada, there are many federal, state, and provincial agencies involved in the management of both the lake and its watershed. The Lake Champlain Basin Program (LCBP), created in 1990, is a partnership organization that coordinates these management entities and that has developed and is now implementing a comprehensive management plan for the lake and surrounding area (Lake Champlain Basin Program 2003). This management plan consists of a number of goals related to issues such as phosphorus pollution; toxic contamination; nonnative nuisance species; human health; fish and wildlife communities; wetland, stream and riparian habitat; recreation; cultural heritage resources; and economic development. Because there are financial and other limits for managing Lake Champlain, and the plan is so broad, there are many competing—and even potentially conflicting—goals. It follows that there are many potential indicators that could be used to help monitor and manage the watershed, and these potential indicators may compete for management attention and may even conflict with one another.

Ecosystem Indicators

To analyze the relative importance of these potential indicators and the tradeoffs among them, a stated choice study, as described in chapter 7, was conducted. Because the number of indicators that can be accommodated in this type of study is limited by statistical complexity, the study was designed to focus on five potential indicators: public beach closures, water clarity, land-use change, fish consumption advisories, and the spread of the invasive water chestnut. These indicators were chosen for several reasons. First, each represents at least one of the goals for the Lake Champlain watershed as identified by the LCBP management plan. For example, water clarity represents the goal of phosphorus pollution control. Second, indicators were selected to be as understandable as possible to the public. Again, the example of water clarity representing the issue of phosphorus pollution is illustrative. Finally, only indicators that were responsive to management were included. For example, the spread of zebra mussels, an issue (halting the spread of nonnative nuisance species) that is identified as a goal of the LCBP plan, was excluded because there is currently no feasible method for managing zebra mussels.

The five potential indicators included in the study were organized into a stated choice study. Each indicator was assigned a range of three standards that could represent future conditions in the watershed. Indicators and standards used in the study are shown in table 19.1. A visual approach (as described in chapter 6) was used to help describe the range of standards for water clarity. For

TABLE 19.1. Ecosystem indicators and standards
used in the Lake Champlain Basin study

Indicator	Standards
Public beach closures	Beaches not closed
	Beaches closed 7 days/year on average
	Beaches closed 14 days/year on average
Water clarity	No algae blooms that produce surface scum
	10 days/year of algae blooms with some surface scum
	10 days/year of algae blooms with thick surface scum
Land-use change	Current land-use distribution
	Increase urban/suburban, decrease agriculture
	Increase urban/suburban, decrease natural
Fish consumption advisories	Safe to eat unlimited fish
	Safe to eat one fish/month
	Not safe to eat fish
Spread of water chestnut	Extent of water chestnut reduced by 10 miles
	Extent of water chestnut not changed
	Extent of water chestnut increased by 10 miles

each indicator variable, one standard represented existing conditions, and the other two standards represented better and worse conditions than might result from more or less intensive management. The three standards of each indicator were arranged into paired comparisons following a fractional, factorial design (Louviere and Timmermans 1990). The resulting choice experiment consisted of five blocks of nine paired comparisons. Each block was assigned to a different version of a survey questionnaire. An example of a paired comparison question is shown in figure 19.1.

The resulting study questionnaire was mailed to a representative sample of Vermont, New York, and Quebec residents who were on the LCBP mailing list. Respondents were instructed to examine and consider the standards of the five indicator variables in the two profiles (*Lake Champlain A* and *Lake Champlain B*) presented in each paired comparison and choose the profile they preferred. The choice task was likened to the tradeoffs that managers have to make when prioritizing management activities and allocating management resources. An example question with instructions was followed by nine, paired comparison-choice questions. Respondents had to tradeoff two, three, or four indicators for each comparison. The questions were ordered such that the simpler, two-indica-tor, tradeoff questions were first, followed by the more complex tradeoffs.

Study data were analyzed by examining the effects of the indicators and standards comprising the profiles on the preferences of the respondents. Regression analysis was used for this purpose, with the choice of profile *A* or *B* being the

1. Do you prefer Lake Champlain A or Lake Champlain B?

? **Lake Champlain A**	? **Lake Champlain B**

Lake Champlain public beaches are **closed for 14 days** per summer on average because bacteria levels exceed local standards.

Lake Champlain public beaches **do not have to be closed** at all during the summer because bacteria levels exceed local standards.

There are **no algae blooms that produce surface scum** in the main lake.

There are **10 days of algae bloom with thick surface scum** in the main lake.

Lake Champlain Basin land use distribution is:

Lake Champlain Basin land use distribution is:

Agricultural 16%

Urban and Suburban 12%

Forests and other natural habitats 72%

Agricultural 16%

Urban and Suburban 12%

Forests and other natural habitats 72%

It is **safe to eat only 1 fish** per month from Lake Champlain due to the toxic substances found in their bodies

It is **safe to eat only 1 fish** per month from Lake Champlain due to the toxic substances found in their bodies

The northward extent of water chestnut (shown in black) is reduced by 10 miles.

The northward extent of water chestnut (shown in black) is increased by 10 miles.

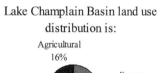

Figure 19.1. Example paired comparison from the choice experiment questionnaire.

dependent variable, and the standards for each indicator being the independent variables. Nearly all of these relationships were statistically significant, meaning that the independent variables (the indicators and standards) explain much of the variation in the dependent variables (the choices that respondents made between the paired comparisons).

The first and third standards for the indicator of safe fish consumption were the strongest predictors of respondent choices between profiles. This implies that safe fish consumption is the most important of the five indicators to the study population. Based on these results, management actions to reduce toxic substances in the water and consequently in the tissue of fish should receive strong consideration compared to management actions addressing the other ecosystem characteristics included in this choice experiment. Algae blooms, beach closures, and the spread of water chestnut attributes were all of similar significance and prediction strength. The land-use attribute was a weaker predictor of choice. Although land use is an important pressure on the lake ecosystem, it is not a characteristic that respondents directly associated with lake use and enjoyment. Therefore, it may be the least important predictor because it was the most abstract and the least directly relevant to respondents' experience of the lake.

Findings for the land-use distribution indicator vary from the pattern of the other indicators because the relationship between the standards of the land-use indicator was not ordinal. Standard one, the current land-use distribution, had a positive and statistically significant coefficient; and the second standard, converting agricultural land into urban or suburban land, had a negative and statistically significant coefficient of approximately the same magnitude. The third standard of the land-use indicator, converting natural land cover to urban/suburban land cover, had a negative coefficient, but it was not a statistically significant predictor of profile choice. Statistics for the standards of this indicator suggest that respondents preferred less land development and that preserving the agricultural landscape was more important than preserving the natural landscape.

Standards for Ecosystem Indicators

The study is illustrative of methods that can be used to engage society in identifying and prioritizing potential indicators for broad-based environmental planning and management. But what are societal standards for these indicator variables? As noted at the beginning of this chapter, this has been a challenging issue for environmental planning and management. A second component of research on the Lake Champlain watershed was designed to help address this issue. This research adopted normative theory and methods, as described in chapter 5, in helping to formulate standards for ecologically related indicators.

Based on the goals and objectives identified in the LCBP management plan, eight potential indicator variables were selected for inclusion in this component of the study. These included the five indicators addressed in the first component of research described earlier, plus the impact of sea lampreys on sport fish, water pollution in tributary streams, and public recreational access to the lake. Personal norms for these eight potential indicators were measured through a series of questions included in the public survey, described earlier. For each indicator, respondents were asked to rate the acceptability of a range of potential standards represented by five or six discrete levels of each indicator variable. Acceptability was measured on an ordinal scale that ranged from −3 ("very unacceptable") to +3 ("very acceptable").

Social norm curves were created by plotting mean acceptability ratings across the ranges of standards examined for each indicator, as shown in figure 19.2. The norm curve for public beach closure had high salience (as described in chapter 5), as indicated by a wide range of values on the y axis. This curve shows that on average, seven or more days of public beach closures during the summer is unacceptable. Though there are some notable exceptions, most beaches on Lake Champlain easily meet the standard. Algae blooms, which also impact recreational use and enjoyment of Lake Champlain, show a similar norm curve shape. Moreover, the shift from acceptable to unacceptable conditions occurs at approximately seven days for both public beach closure and days of intense algae blooms. It should be noted that the questions addressing these indicators did not appear in the same questionnaire. To reduce respondent burden, several versions of the study questionnaire were used, and each version included normative questions about only two indicator variables.

The norm curve for safe fish consumption showed less salience than the curves generated for other indicators. However, the range of potential standards presented in the question for this indicator was more constrained than the ranges used for other indicators. Current fish consumption advisories suggest that adults limit their intake of lake trout and walleye caught in Lake Champlain to about one meal per month. (The advisories are more restrictive for women of childbearing age and children and less restrictive with regard to limiting the intake of other fish species caught in the lake.) Nevertheless, there appears to be little public tolerance for limitations on safe fish consumption, based on study findings.

There are currently at least twenty-seven streams impaired by storm-water pollution in the Lake Champlain watershed. This is well beyond the minimum acceptable condition based on the social norm curve for impaired streams. In this curve, the mean ratings for all potential standards except the first are in the unacceptable range, and crystallization around the means increases steadily for each level beyond nine impaired streams. At twenty-seven impaired streams (the current condition), the standard deviation is well below the minimum acceptable condition. This suggests that management attention is necessary to bring stream impairment in line with societal norms.

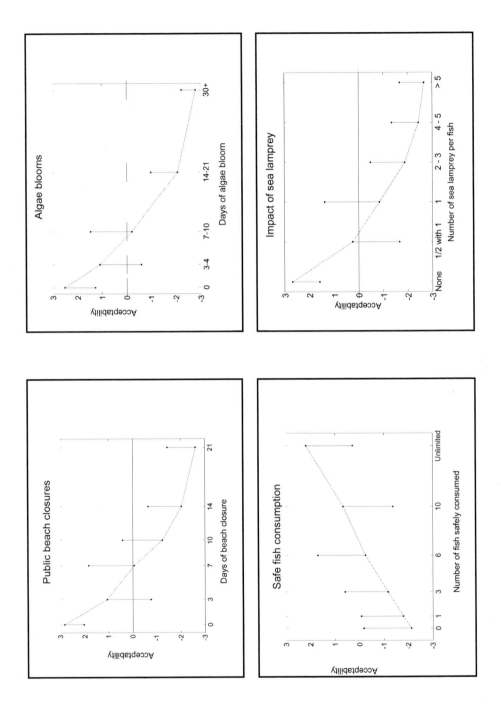

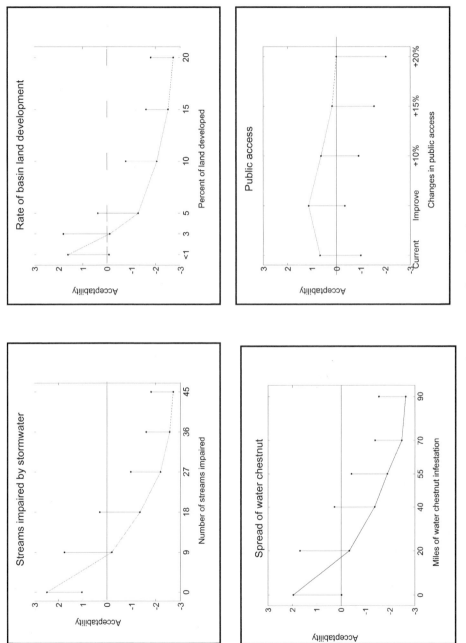

Figure 19.2. Social norm curves for study indicators.

The social norm curves for land development and stream impairment have a similar pattern with decreasing variability about the means in the unacceptable range. To measure the social norm curve for land development, respondents were told that there were at least 300,000 acres of developed land in the Lake Champlain watershed. They were then asked to rate the acceptability of a range of increases in developed land per year. The results clearly indicate that increasing land development is not favored. Rates of land development vary across the watershed. While growth in Chittenden County, Vermont, may be outside the range of acceptability, rates of land development are lower in other areas of the watershed, particularly in New York.

Currently, water chestnut in Lake Champlain extends about fifty-five miles north of Whitehall, New York, and this condition is not generally acceptable, according to study findings. The LCBP currently has an aggressive program for water chestnut harvesting, which is successfully moving this extent southward. The norm curve suggests that the public would like to see the extent of water chestnut halved and therefore supports continuing this management action.

The social norm curve for sea lamprey wounds shows that one or more wounds on fish caught from Lake Champlain was generally not acceptable. Currently, wounding rates vary considerably by fish species, fish size, and lake segment, but most fish have at least one wound. To achieve a lower level of lamprey wounding, a higher level of lamprey control will be needed.

Both the sea lamprey and water chestnut indicator norm curves support continued management of these nuisance species. However, written comments qualitatively demonstrate concern over the use of chemicals in the treatment of nuisance species. Several respondents commented that they were unaware of the water chestnut problem, and others stated that they would not support management to reduce water chestnut if it involved the application of chemicals. There was also concern expressed about the use of lampricides in the watershed. These results are consistent with the findings for safe fish consumption and in sum suggest that there are concerns about the release and presence of toxic chemicals in Lake Champlain, no matter what their source or purpose.

The final norm curve generated was for public access. This curve shows the smallest range in acceptability ratings and is the most difficult to interpret. Although improving public access was identified as the optimal management action, there was little difference between any of the responses to this question. This seems surprising because public access is generally considered an important issue for Lake Champlain. In a separate question, respondents were asked to rate the importance of public access, and 82% of respondents indicated that public access was "very important" or "important." Public access was also commonly mentioned in the comment section of the questionnaire.

The reasons why the norm curve results and other survey comments about public access may appear inconsistent are not clear. Because the potential standards listed as choices in this question were not ordinal, as with the other batteries of normative questions, this question may have been more difficult for respondents to answer. Another potential explanation is that the ability to access the lake is highly valued (hence the high importance ranking), but there is general satisfaction with the current level of public access (as shown by the social norm curve). The LCBP has been active in both improving existing access and developing additional access sites throughout its tenure. Study findings, including the social norm curve, support continuing this management action, but it may not be a high priority issue because current conditions are judged by the public to be satisfactory.

Findings from the program of research have been reported to the LCBP. This information will be incorporated into the public planning process for the Lake Champlain Basin to identify the highest priority indicators and to formulate standards for these indicator variables. These indicators and standards will be monitored on a regular basis and will become the focus of an annual "report card" on the ecological and related social status of Lake Champlain and the surrounding area and will help guide environmental management in this area.

Conclusion

We can never do nothing.

How much can we use the environment without spoiling it? This is most fundamental question in all of environmental management. We have wrestled with this question across the span of human history, and it is now firmly on our environmental agenda in the form of common property resources, carrying capacity, and the emerging notion of sustainability. Manifestations and applications of this question continue to grow in number, scale, and urgency. This book has addressed how this question applies to parks and protected areas, with special attention to the U.S. national park system. The National Park Organic Act of 1916 is a classic example of the inherent tension between using the environment and protecting what we find so valuable about it: national parks and related areas must meet both of these often competing demands.

Parks and protected areas are also examples of common property resources and subject to the tragedy of the commons as described so hauntingly by Garrett Hardin. His original paper in *Science*, with which this book began, is now a foundational element of modern environmental thought. Common property resources are especially vulnerable to overexploitation because the environmental (and related social) degradation caused by their use is not borne fully by individual users (but is borne by society at large). There is a built-in incentive for "rational" individuals to overexploit common property resources. In the case of parks and protected areas, people will continue to visit these areas because—at the level of the individual—the benefits they receive outweigh the costs they must pay. However, at the greater societal level, the parks are nonetheless degraded and their value is ultimately diminished.

Concern over the tragedy of the commons—and environmental management more broadly—is driven by an assumption that there are limits to our use of the environment. This issue is often considered within the rubric of carrying capacity. The concept of carrying capacity has a relatively long and rich history in environmental and natural resource management. In its broadest manifestation, it is applied to the population of humans that can ultimately be accommodated in a given area or even on the planet as a whole. Initial scientific applications of carrying capacity (such as the logistic growth model) suggested that populations will grow until they are constrained by environmental factors.

More recent thinking suggests that both the tragedy of the commons and carrying capacity are less deterministic and more normative than originally conceived. For example, the rigid notion of rationality assumed in the tragedy of the commons might be tempered by some degree of altruism or enlightened self-interest on the part of individuals to protect what is important to society as a whole (and to individuals as well, at least in the long run). Moreover, human population growth might ultimately be mindful of both the material and non-material conditions under which people must live (e.g., economic well-being, level of environmental quality). These interpretations suggest that there are (or at least can be) social values and related norms that can guide environmental management, including our use of common property resources and carrying capacity. Such norms can be the basis of "mutual coercion, mutually agreed upon," the social action that Hardin suggests is required to resolve the tragedy of the commons, carrying capacity, and related environmental issues.

This shift in thinking has led to a new emphasis on defining the environmental and related conditions that society wishes to maintain. There is intuitive and growing scientific understanding that increasing population and associated economic growth can lead to an array of environmental impacts and related social costs. The operative questions associated with carrying capacity, managing common property resources, and environmental management more broadly then become (1) what levels of environmental impacts are acceptable? and (2) what type of environmental and related social conditions do we want to maintain?

Contemporary application of carrying capacity to parks and related areas has followed (sometimes even led) this line of thinking. The concept of "limits of acceptable change" was introduced into the parks and outdoor recreation literature several decades ago. Moreover, it has become clear that there cannot be any one inherent carrying capacity of a park or related area. Rather, carrying capacity can be determined only as it relates to environmental and associated social objectives. For example, what degree of environmental protection should be maintained, and what type of visitor experience should be provided? These types of management objectives are sometimes called *desired conditions*. Moreover, management objectives/desired conditions should ultimately be expressed in quantitative terms—generally called *indicators* and *standards*—so carrying capacity can be measured empirically. Indicators are measurable, manageable variables that help define the quality of parks and related areas; they are proxies of management objectives/desired conditions. Standards define the minimum acceptable condition of indicator variables. Based on this conceptual framework and terminology, carrying capacity of parks and protected areas can then be defined in an operational way as the level and type of visitor use that can be accommodated without violating standards for relevant indicator variables.

This approach to carrying capacity has been designed into several planning/management frameworks for parks and related areas. The most recent framework is Visitor Experience and Resource Protection (VERP) developed by the National Park Service (1997) (Manning 2001), and this framework has been used in this book to illustrate recent applications of carrying capacity to parks and related areas. However, all of the contemporary carrying capacity frameworks are built upon the conceptual foundation described earlier and function through a similar core sequence of steps (Manning 2004):

1. Establish management objectives/desired conditions and associated indicators and standards.
2. Monitor indicator variables.
3. Apply management practices to ensure that standards are maintained.

There are important corollaries to these developments in thinking about the carrying capacity of parks and protected areas. As applied to parks and related areas, carrying capacity has resource, experiential, and managerial components. Management objectives/desired conditions and associated indicators and standards should be considered for each of these components. Carrying capacity also has descriptive and prescriptive components. The descriptive component addresses the relationships between (1) levels and types of visitor use, and (2) resulting impacts to park resources, experiences, and management. The prescriptive component addresses the seemingly more subjective issue of how much impact (and, therefore, how much use) should be allowed. As might be expected, the prescriptive component of carrying capacity is often the most challenging. Finally, carrying capacity determination will always require some element of management judgment; if alternative carrying capacities are possible, then some judgment will have to rendered as to which is the most appropriate. However, such judgments can and should be informed by research.

In keeping with these ideas, application of carrying capacity to parks and related areas is now being informed by both natural and social science research, and these research approaches are beginning to be integrated where appropriate. Carrying capacity is an inherently interdisciplinary concept (as is environmental management more broadly), and interdisciplinary research will ultimately be needed. A program of natural science–based research on the ecological impacts of outdoor recreation—often called *recreation ecology*—has generated an increasing body of knowledge that can be used to help formulate indicators and associated standards for parks and related areas. A program of social science–based research has also evolved that includes a variety of theoretical and methodological approaches and a growing body of knowledge about the experiential impacts of outdoor recreation and that can also help formulate indicators and standards. Taken together, this research can help address the resource, experiential, and managerial components of carrying capacity. And, where needed, this research can be integrated to address the inevitable nexus among these components. For example, at what point do resource-related impacts of recreation degrade the quality of the visitor experience, how can visitors be encouraged to mitigate their environmental and experiential impacts, and how acceptable are management actions in light of desired visitor experiences?

Although natural science–based research has been addressed, this book has emphasized social science–based research for two reasons. First, the recreation ecology literature has been synthesized and reported effectively in recent texts and reports (e.g., Hammitt and Cole 1998; Leung and Marion 2000). Second, there is a growing notion that carrying capacity may be largely a social issue driven by the needs and wants of society. In fact, a recent report on carrying capacity in its application to broad issues of environmental management suggests that carrying capacity might really be *social* carrying capacity (Seidl and Tisdell 1999). This may be overstated; it seems foolish to deny that there are environmental constraints to the carrying capacity of parks or in any other context. However, these constraints might often be wide ranging. In such cases, society will play a vital role in determining carrying capacity, and social science (integrated with natural science where appropriate) will facilitate this process.

A range of social science research methods have been adapted and applied to carrying capacity analysis and management. Perhaps the most important is normative theory and methods. If carrying capacity is a normative rather than deterministic concept as is suggested by recent thinking, then social norms are at the heart of measuring and managing carrying capacity. Similarly, social norms must ultimately guide the "mutual coercion, mutually agreed upon" needed to manage parks and other common property resources. Research suggests that visitors and other interested publics have normative standards about appropriate environmental, experiential, and managerial conditions in parks and related areas. Moreover, these norms can be measured and can help inform the development of management objectives/desired conditions and associated indicators and standards. Visual research methods can be used in selected contexts to help make normative research more valid and effective. Several forms of tradeoff analysis can be used to help ensure that normative questions and the answers they elicit are as informed as possible about potentially competing and even conflicting objectives. Qualitative and quantitative survey research can help identify the most important or salient societal indicators of resource, experiential, and managerial conditions in parks and related areas. Computer simulation modeling of visitor use can help develop important baseline data on visitor-use levels and patterns (and resulting environmental and social impacts); can help monitor indicator variables that are inherently difficult to observe; can help estimate carrying capacity based on crowding-related concerns; and can help predict the potential effectiveness of alternative management practices.

The case studies outlined in this book are examples of the ways in which these research methods are being applied and carrying capacity is being measured, analyzed, and managed in a variety of resource, cultural, and social contexts across the national park system. A range of indicators of the resource, experiential, and managerial components of carrying capacity are being identified, and associated standards are being formulated. Indicators are being monitored and management actions taken to ensure that standards are maintained. This constitutes measurement and management of carrying capacity. In the special issue of *Science* in 2003, celebrating the 35th anniversary of the publication of Hardin's "The Tragedy of the Commons," the editor-in-chief, Donald Kennedy, wrote in his introduction that in our efforts to address management of common property resources, carrying capacity, and sustainability, "there have been some real winners, such as managed preserves that blend conservation objectives with recreational values" (Kennedy 2003, 1861). This is hopeful.

Efforts to address carrying capacity would ring hollow without feasible and effective management practices. Fortunately, there is a range of possibilities for parks and protected areas. Of course, we can increase the number and/or size of such areas to accommodate more visitors. When this option is not available, we can limit demand through restrictions on the amount of use. Or we can educate visitors in ways that will limit their environmental and social impacts. Where management objectives allow, we can harden resources to impact through a variety of site management practices. In the context of broader environmental management, analogous options have been described as the "bigger pie," "fewer forks," and "better manners" schools of management (Cohen 1999). It is important to consider the full range of management alternatives. Research in parks and outdoor recreation has begun to assess the potential effectiveness of alternative management practices, and some of the findings are encouraging. But more research is needed.

This book addresses carrying capacity primarily in the context of parks and protected areas. But the concepts, principles, and approaches might be equally applicable to much of the broader field of environmental management. Both carrying capacity and the newer concept of ecosystem management stress the relationships between the environment and society that must be addressed. There are obvious parallels between carrying capacity and the emerging concept of sustainability as both address the inherent tension between use of the environment and protection of its basic integrity. Indicators of environmental and related social conditions are a cornerstone of contemporary carrying capacity frameworks, and for similar reasons and in similar ways indicators are becoming a cornerstone of environmental management in many of its applied fields. Associated standards for indicator variables are receiving needed attention in the field of parks and related areas but need more work in broader environmental management if the concept of indicators is to be fully useful. Adaptive management is growing in importance as a guiding concept in environmental management and is inherent in carrying capacity frameworks for parks and related areas as management actions are guided by monitoring of indicator variables.

Armed with a conceptual foundation and related set of terminology, an associated planning/management framework, a growing set of supporting research approaches, an array of management alternatives, and a number of hopeful case studies, we are ready to engage the carrying capacity of parks and protected areas more deliberately. Of course, applying these research, planning, and management tools will be challenging and sometimes even contentious. But failure to do so will be even more painful in the long run. Do we want to conduct our management of parks and protected areas—the crown jewels of our natural and cultural heritage—by design or by default? As Hardin (1968) suggests, "We can never do nothing." If we choose not to manage the carrying capacity of parks and protected areas, we are implicitly deciding that their current conditions are acceptable and that trends in use and related impacts are not worrisome. We should find comfort and courage in the democratic and civic character of the substance and process outlined in this book. Management of parks and protected areas—and of broader environmental issues—should be based on societal values and related norms, not on privilege bestowed by power or even scientific knowledge. Engaging the public in decisions about managing parks and protected areas builds trust, ownership, and the "social capital" that engenders public enthusiasm and support (Minteer and Manning 2003; LaChappelle and McCool 2005; Manning and Ginger, forthcoming). While this process of engagement can be challenging even at local levels, it can appear especially daunting at larger scales (Manning and Ginger, forthcoming). Parks and protected areas often have national and even international significance. And environmental management more broadly must increasingly be conducted at a global level. Some of the research methods outlined in this book, especially those that employ representative sampling approaches, can be useful at these higher scales. Management decisions that inherently limit personal freedoms (at least in the short run) are likely to be contentious, but they are more apt to endure if they are built upon the values and norms of those they most directly affect.

Despite advances in theory and related empirical methods, some measure of management judgment will remain inescapable. However, when this judgment is rendered in the context of a rational, transparent, conceptual and planning framework, and when it is supported by informed research and related public engagement, it will lead to a program of management that protects both the environment and the public good. If freedom is truly the recognition of necessity, then it is time to move ahead in the management of parks and protected areas and the broader field of environmental management.

Appendix A

Indicators for Parks and Protected Areas

Study	Area	Respondents	Indicator
Mergliano 1990	Wilderness	Wilderness managers and scientists	• Number of campsites above an acceptable impact index • Percentage of visitors who report seeing wildlife • Range condition and trend • Air visibility—extinction coefficient or visual range • Litter quantity—number of pieces of litter/campsite or/trail mile; number of pounds of garbage packed out each season • Number of manager-created structures • Number of signs/trail mile • Trail condition—length of multiple trails or number of trail miles with unacceptable problems to visitors (e.g., depth exceeding eight inches, year-round muddiness) • Length of trail in areas managed as trailless • Fecal coliform/fecal streptococci ratio (drinking water quality) • Number of occupied campsites within sight or sound of each other or visitor report of number of groups camped within sight or sound • Number of violations of no-trace regulations • Percentage of groups carrying a stove (not using a campfire)

Study	Area	Respondents	Indicator
Mergliano 1990 *cont'd*			• Number of occurrences of unburied human feces • Number of occurrences of motorized noise/day • Percentage of the season wilderness rangers are out patrolling the area • Number of regulations that limit visitor use or restrict travel • Number of regulatory signs posted beyond trailhead
Shindler and Shelby 1992	Wilderness campsites	Members of five interest groups	• Amount of bare ground • Size and appearance of fire rings • Distance from trail • Screening from other sites • Out of sight/sound of other sites • Evidence of litter • View of scenery • Available firewood • Sheltered from weather • Dry and well drained • Water for aesthetic reasons • Flat place for sleeping • Close to good fishing • Logs and rocks for seating • Close to drinking/cooking water
Whittaker 1992	Five Alaska rivers	Floaters, motor boaters	• Litter • Signs of use • Campsite competition • Fishing competition • Launch congestion • River encounters • Camp encounters • Powerboat use • Airboat use • Rafting/canoeing use • Airplane landings • Helicopter landings • Off-road vehicle use • Hazard signs • Interpretive signs • Public-use cabins • Private cabins • Concessions • Long-term camps

Study	Area	Respondents	Indicator
Roggenbuck et al. 1993	Four wilderness areas	Visitors	• Amount of litter I see • Number of trees around campsite that have been damaged by people • Amount of noise associated with human activities within the wilderness • Amount of human-made noise originating from outside the wilderness • Number of wild animals I see • Amount of vegetation loss and bare ground around a campsite • Number of horse groups that camp within sight or sound of my campsite • Number of hiker groups that camp within sight or sound of my campsite • Number of horse groups that travel past my campsite while I am there • Number of campfire rings that people have made • Number of hiker groups that walk past my campsite • Number of large groups that I see along the trails • Number of horse groups I see along the trails in a day • Percentage of time other people are in sight when I'm on the trail • Visibility of lights originating from outside the wilderness • Total number of people I see hiking along the trail • Number of groups of hikers I see along the trail • Amount of time I spend traveling on old roads in the wilderness • Number of miles of gravel road I travel to get to the wilderness
Shafer and Hammitt 1994	Cohutta Wilderness, GA	Visitors	• The total amount of time that your party has in an area without seeing or hearing anyone else • The amount of restriction management places on where you may travel in the area • The number of permanent structures placed by management in the wilderness

Study	*Area*	*Respondents*	*Indicator*
Shafer and Hammitt 1994 *cont'd*			• Seeing an unusual type of plant • The amount of restriction management places on where you may camp in an area • The level of difficulty required to obtain an overnight permit • The number of vehicles you see at the trailhead • The number of fire rings found in a campsite • The number of days in a row you are able to stay in the wilderness on a given trip • The number of signs designating locations in the wilderness • The number of groups you pass during the day while traveling • Having signs placed by wilderness managers that state regulations about wilderness • The amount of wilderness that does not have trails in it • The distance of campfires from trailheads • The number of rangers you see in the area • The amount of ranger contact in the backcountry to check your permit and/or explain regulations about use • The amount of litter found in campsites • The amount of litter seen along the trail • The number of trees or other vegetation damaged by previous users • The amount of noise heard in the area that comes from outside the wilderness • The amount of fully mature forest in the wilderness area • Observing a natural ecosystem at work • The amount of solitude your group experiences • The amount of noise heard in the area that comes from other wilderness visitors

Study	Area	Respondents	Indicator
			• The number of different species of wildlife you see • The number of areas in the wilderness that are very remote • The distance between your campsite and the campsite of others • Seeing specific types of wildlife • The amount of light visible at night that comes from outside the wilderness • The level of trail maintenance • The number of groups that pass within sight of your camp • An area in the wilderness that is left completely primitive (no trails, bridges) • Having a portion of the wilderness where camping location is unconfined • Having trail markers placed by management (blazes, cairns, posts)
Manning et al. 1995b, 1995c, 1996b; Manning and Lime 1996	Arches National Park, UT	Visitors	• Orientation, information, and interpretive services • Number and type of visitor facilities • Number of people encountered • Visitor behavior and activities • Resource impacts • Park management activities • Quality and condition of natural features
Jacobi et al. 1996	Acadia National Park, ME	Carriage road visitors	• Number of visitors encountered • Type of visitors encountered (hikers or bikers) • Behavior of visitors (speed of bikers, keeping to the right, obstructing the roads, traveling off the roads)
Lime et al. 2001	Arches National Park, UT	Visitors	• PAOT at attraction sites
Manning et al. 1997b	Grand Canyon National Park, AZ	Day Hikers	• PPV on different trail types in different seasons

Study	Area	Respondents	Indicator
Lime et al. 2001	Arches National Park, UT	Visitors	• PAOT at attraction sites
Manning et al. 1998b	Alcatraz Island, CA	Visitors	• PAOT in the cellhouse
Manning et al. 1998c	Yosemite National Park, CA	Visitors	• PAOT at attraction sites • PPV on trails
Warzecha et al.	Arches National Park, UT; Capital Reef National Park, UT; Colorado National Monument, CO; Natural Bridges National Monument, UT	Visitors	• PPV on trails
Manning et al. 1999c	Statue of Liberty/Ellis Island National Monument, NY, NJ	Visitors	• Waiting time to get into the statue • Waiting time to buy a ferry ticket • Waiting time to board the ferry
Manning et al. 1998c	Yosemite National Park, CA	Visitors	• PAOT at attraction sites • PPV on trails
Borrie et al. 1999	Yellowstone National Park, WY MT, ID	Winter visitors	• Snowmobiles encountered/hour
Vande Kamp et al. 2003	Kenai Fjords National Park, AK	Visitors	• PAOT at attraction site
Manning et al. 2002c	Mesa Verde National Park, CO	Visitors	• PAOT at attraction sites • PPV on trail • Tour frequency • Tour group size

Study	Area	Respondents	Indicator
			• Tour length
			• Number of tour groups at one time
Manning et al. 2002d	Acadia National Park— Schoodic Peninsula, ME	Visitors	• Trail impacts • VPV on the scenic road • PAOT at attraction sites
Manning et al. 2002e	Blue Ridge Parkway, VA/NC	Visitors	• VPV on the road • PAOT at scenic overlooks
Park Studies Lab 2003	Hawaii Volcanos National Park, HI	Visitors	• PAOT at attraction sites • Cars parked on road • PPV on trails • Walking time from parking area to attraction site
Bacon et al. 2004	Acadia National Park—Isle au Haut, ME	Visitors and residents	• Encounters/day • Trail impacts • Visitor-caused trails • Trail development
Manning et al. 2004b	Zion National Park, UT	Visitors (2002, 2003)	• Group size • Number of hikers seen/day • Number of groups of hikers seen or heard/day • Trail impacts • Campsite impacts • Trail development • Campsite development
Manning and Budruk 2004	Boston Harbor Islands, MA	Visitors	• Hikers seen/hour • Trail impacts • Campsite impacts • Litter • Frequency of ferry service • Frequency of water shuttle service • PAOT at attraction sites
Park Studies Lab 2004	Haleakala National Park, HI	Visitors	• PAOT at attraction sites

Note: The National Park Service has compiled a list of indicators that have been used in carrying capacity and related plans for a variety of parks and protected areas. This information can be found at the following website: www.usercapacity.nps.gov/is.

PAOT = people-at-one-time; PPV = persons/viewscape; VPV = vehicles/viewscape.

Source: Manning, R., M. Budruk, W. Valliere, and J. Hallo. 2005. *Research to Support Visitor Management at Muir Woods National Monument and Muir Beach: Study Completion Report.* Burlington: University of Vermont, Park Studies Lab.

Appendix B

Standards for Parks and Protected Areas

Study	Area	Respondents	Indicator	Normative standard	
				Mean	Median
Stanley 1973	Boundary Waters Canoe Area, MN	Visitors	• Encounters with paddling canoeists		3.5
			• Encounters with motor canoeists		0.0
			• Encounters with motorboats		0.0
	Three wilderness areas	Visitors	• Encounters with back-packing parties		2.5
			• Encounters with horse parties		1.8
Stanley 1980a	Desolation Wilderness, CA	Visitors	• Encounters with back-packing parties		9.5
			• Encounters with horse parties		4.0
			• Encounters with large parties		2.6
			• Parties camped within sight or sound		2.4
	Spanish Peaks Wilderness, MT	Visitors	• Encounters with back-packing parties		4.5
			• Encounters with horse parties		3.5
			• Encounters with large parties		1.8
			• Parties camped within sight or sound		1.9

Study	Area	Respondents	Indicator	Normative standard Mean	Normative standard Median
Shelby 1981	Three rivers	Boaters	Colorado River, Grand Canyon National Park, AZ		
			• Encounters/day		0.9/2.4/4.0 [a]
			• Hours in sight of others each day		0.5/0.7/1.5
			• Number of stops out of ten with encounters		0.7/2.0/3.8
			• Chances of meeting twenty to thirty people at popular places on the river		9%/23%/ 41%
			• Number of nights out of ten camped near others	0/1.3/3.0	
			Rogue River, OR		
			• Encounters/day		1.5/2.9/4.4
			• Hours in sight of others each day		0.5/1.0/1.9
			• Number of stops out of five with encounters		0.6/1.6/2.3
			• Chances of meeting five to twenty people at popular places on the river		12%/28%/ 44%
			• Number of nights out of five camped near others		0/1.1/2.1
			Illinois River, OR		
			• Encounters/day		0.7/2.0/2.7
			• Hours in sight of others each day		0.4/.9/1.6
			• Number of stops out of five with encounters		0.2/1.3/1.8
			• Number of nights out or three camped near others		0/0.2/0.7
Heberlein et al. 1986	Apostle Islands National Lakeshore, WI	Boaters	• Number of boats moored at Anderson Bay		11.0
			• Number of boats moored at Quarry Bay		11.0
Vaske et al. 1986	Brule River, WI	Floaters	• Encounters with fishermen	7.2	
			• Encounters with canoers	5.7	
			• Encounters with tubers	2.3	

				Normative standard	
Study	*Area*	*Respondents*	*Indicator*	*Mean*	*Median*
Shelby et al. 1988a	Rogue River, OR	Boaters	• Encounters/day on river	5.7	
			• Number of nights out of five camped near others	1.4	
Shelby et al. 1988b	Mt. Jefferson Wilderness, OR	Campers	Maximum size of fire rings		
			• Hunts Lake	20 inches	
			• Russell Lake	34 inches	
			Maximum area of bare ground		
			• Hunts Lake	750 sq. ft.	
			• Bays Lake	750 sq. ft.	
			• Scout Lake	1,450 sq. ft.	
Whittaker and Shelby 1988	Deschutes River, OR	Boaters	• Hours in sight out of four		1.8–2.2[b]
			• Incidents of discourteous behavior/day		0.1–0.2
			• Number of stops out of four where human waste is seen		0.1–0.3
			• Jet boats encountered/day		0.3–1.3
			• Boats/hour passing anglers		4.0–4.7
			• Fishing holes passed up out of four due to competition		1.3–1.7
			• Minutes waiting to launch		10.3–14.9
			• Nights out of four camped with other groups		1.4–1.9
			• Nights out of four camped near other groups		0.4–0.9
			• Camps passed up out of four due to competition		1.1–1.2
			• Camps out of four with fire rings present		0.5–1.1
Patterson and Hammitt 1990	Great Smoky Mountains National Park, NC,TN	Backpackers	• Encounters at trailhead	3.9	3.0
			• Encounters on trail	5.5	4.0
			• Encounters at campsite	2.7	2.0
Roggenbuck et al. 1991	New River, WV	Floaters	Number of boats seen		
			• Wilderness whitewater	10.1	
			• Scenic whitewater	20.4	
			• Social recreation	33.4	
			Percentage of time in sight of other boats		
			• Wilderness whitewater	18.3	
			• Scenic whitewater	32.3	
			• Social recreation	48.1	

Study	Area	Respondents	Indicator	Normative standard	
				Mean	Median
Roggenbuck et al. 1991 *cont'd*			Number of rapids having to wait		
			• Wilderness whitewater	1.2	
			• Scenic whitewater	2.4	
			• Social recreation	4.0	
Young et al. 1991	Cohutta Wilderness, GA	Visitors	• Number of people hiking on trail in a day	11.5	
			• Number of large groups hiking on trail a day	3.4	
			• Number of hiker groups camped in sight or sound of campsite	2.2	
			• Number of hiker groups walking past campsite	3.7	
			• Number of horse groups seen on trail in a day	2.4	
			• Number of horse groups camped in sight or sound of campsite	1.7	
			• Percentage of time other people are in sight while on trail	13.9	
			• Number of groups of hikers seen on trail in a day	3.9	
			• Number of horse groups that travel past my campsite	1.2	
Martinson and Shelby 1992	Three rivers	Salmon anglers	Encounters with bank anglers *Preferred*		
			• Klamath		—
			• Waimakariri		3.6
			• Lower Rakaia		3.5
			• Upper Rakaia		<1.0
			Tolerable		
			• Klamath		12.6
			• Waimakariri		6.9
			• Lower Rakaia		9.5
			• Upper Rakaia		3.8
Shelby et al. 1992	Colorado River, Grand Canyon National Park, AZ	Guides and trip leaders	• Minimum stream flow	10,000 cfs	
			• Maximum stream flow	45,000– 50,000 cfs	
Williams et al. 1992a	Four wilderness areas	Visitors	• Encounters with hiking groups along trail	8.7–11.6[c]	
			• Encounters with horse groups along trail	5.1–6.4	
			• Encounters with large groups along trail	5.8–7.1	
			• Hiker groups camped within sight or sound	3.8–6.9	

Study	Area	Respondents	Indicator	Normative standard	
				Mean	*Median*
			• Horse groups camped within sight or sound	3.1–3.8	
			• Hiker groups passing by camp	5.5–7.9	
			• Horse groups passing by camp	5.4–7.4	
Roggenbuck et al. 1993	Four wilderness areas	Visitors	• Number of pieces of litter I can see from my campsite		0–2[d]
			• Percentage of trees around a campsite that have been damaged by people		0–5
			• Number of horse groups that camp within sight or sound of my campsite		1–2
			• Number of hiker groups that camp within sight or sound of my campsite		3
			• Number of large groups (more than six people) that I see along the trail		3–5
			• Percentage of vegetation loss and bare ground around the campsite		10–20
Ewert and Hood 1995	San Gorgonio Wilderness, CA John Muir Wilderness, CA	Visitors	Encounters/day • For urban-proximate wilderness	9.0	
			• For urban-distant wilderness	7.7	
Hammitt and Rutlin 1995	Ellicott Rock Wilderness, SC, NC, GA	Visitors	Encounters at trailhead • Ideal	3.8	
			• Maximum	8.7	
			Encounters on trail • Ideal	3.2	
			• Maximum	6.6	
			Encounters at destination site • Ideal	1.0	
			• Maximum	2.5	
			Encounters at all three sites combined • Ideal	2.7	
			• Maximum	5.9	
Shelby and Whittaker 1995	Dolores River, CO	Boaters	Maximum stream flow • Large rafts	~900 cfs	
			• Small rafts	~750 cfs	
			• Canoes	~300 cfs	
			• Kayaks	~900 cfs	

Study	Area	Respondents	Indicator	Normative standard Mean	Median
Shindler and Shelby 1995	Rogue River, OR	Boaters	Encounters with float parties		
			• 1977	5.7	
			• 1991	7.4	
			Encounters with jet boats		
			• 1977	1.5	
			• 1991	1.5	
			Hours in sight of other parties		
			• 1977	1.3	
			• 1991	1.4	
			Acceptable number of stops out of five to meet another group		
			• 1977	1.8	
			• 1991	1.8	
			Acceptable number of nights out of five to camp within sight or sound of another party		
			• 1977	1.4	
			• 1991	1.2	
Watson 1995	Boundary Waters Canoe Area, MN	Canoers	• Encounters with paddling groups	5.8–8.5[e]	
			• Number of nearby campers	2.5–5.7	
Hall and Shelby 1996	Eagle Cap Wilderness, OR	Visitors	• Encounters with other groups	5.6	4.0
Hall et al. 1996	Clackamus River, OR	Floaters	• Encounters with other boaters	7.5/10.4[f]	6/8
			• Percentage of time in sight of other boaters	49.4/46.6	50/50
			• Number of minutes waiting at launch	16.1/18.1	15/15
Lewis et al. 1996b	Boundary Waters Canoe Area, MN	Canoeists	• Encounters with canoe parties on periphery lakes and rivers	5.1	3.1
			• Encounters with canoe parties on interior lakes and rivers	3.8	2.5
			• Encounters with canoe parties on all lakes and rivers	4.2	2.6
Manning et al. 1996a; 1995b; 1996b; 1995a; Manning and Lime 1996	Arches National Park, UT	Visitors	• PAOT at Delicate Arch	28	
			• PAOT at North Window	20	

Study	Area	Respondents	Indicator	Normative standard Mean	Median
Vaske et al. 1996	Columbia Ice Field, Jasper National Park, Canada	Snowcoach riders and hikers	PAOT at attraction site for snowcoach riders		
			• Canadian	96.2	
			• Anglo-American	100.5	
			• Japanese	114.6	
			• German	104.4	
			• British	84.5	
			PAOT at attraction site for hikers		
			• Canadian	47.3	
			• Anglo-American	55.6	
			• German	42.1	
			• British	41.3	
Manning et al. 1997a	Acadia National Park, ME	Carriage road users	PPV[g] VISUAL APPROACH *Long form*		
			• Hikers only	17	
			• Bikers only	12	
			• Even distribution of hikers and bikers	14	
			Short form		
			• Acceptability	11	
			• Tolerance	25	
			• Acceptability for "others"	15	
			• Management action	18	
			NUMERICAL APPROACH		
			• Hikers only	16	
			• Bikers only	13	
			• Even distribution of hikers and bikers	18	
Tarrant et al. 1997	Nantehala River, NC	Floaters	Maximum encounters tolerable RAFTERS *With rafts*		
			• On the river	28.4	
			• At put-in	12.3	
			• At rapids	9.3	
			With kayaks/canoes		
			• On the river	18.4	
			• At put-in	9.2	
			• At rapids	6.8	
			KAYAKERS/CANOERS *With rafts*		
			• On the river	37.4	
			• At put-in	14.1	
			• At rapids	10.3	
			With kayaks/canoes		
			• On the river	39.9	
			• At put-in	15.5	
			• At rapids	12.1	

Study	Area	Respondents	Indicator	Normative standard	
				Mean	Median
Lime et al. 2001	Arches National Park, UT	Visitors	PAOT at attraction sites		
			Long form		
			• Delicate Arch	37	
			• Windows	23	
			• Devil's Garden	13	
			Short form		
			• Delicate Arch	32.8	36.0
			• Windows	23.1	16.0
			• Devil's Garden	21.2	16.0
Manning et al. 1997b	Grand Canyon National Park, AZ	Day Hikers	PPV on different trail types in different seasons		
			LONG FORM		
			Summer		
			• Threshold trails	5	
			• Corridor trails	9	
			• Rim trails	10	
			Fall		
			• Threshold trails	4	
			• Corridor trails	8	
			• Rim trails	11	
			SHORT FORM		
			Summer		
			• Threshold trails	3.6	4.0
			• Corridor trails	6.9	6.0
			• Rim trails	6.9	6.0
			Fall		
			• Threshold trails	3.5	4.0
			• Corridor trails	7.4	6.0
			• Rim trails	8.3	6.0
Lime et al. 2001	Arches National Park, UT	Visitors	PAOT at Delicate Arch using different methods		
			Long form	35.8	
			Short form	27.3	
			Dichotomous choice	31.8	
Manning et al. 1998b	Alcatraz Island, CA	Visitors	PAOT in the cellhouse		
			Long form	44.0	
			Short form	36.0	
Manning et al. 1998c	Yosemite National Park	Visitors	PAOT at attraction sites		
			Long form		
			• Base of Yosemite Falls	92	
			Short form		
			• Base of Yosemite Falls	75	
			PPV on trails		
			Long form		
			• Trail to Yosemite Falls	40	
			• Trail to Vernal Falls	26	
			Short form		
			• Trail to Yosemite Falls	32	
			• Trail to Vernal Falls	20.6	

				Normative standard	
Study	*Area*	*Respondents*	*Indicator*	*Mean*	*Median*
Warzecha et al. 2001	Arches National Park, UT; Capital Reef National Park, UT; Colorado National Monument, CO; Natural Bridges National Monument, UT	Visitors	PPV on trails Devil's Garden Trail (Arches) Hickman Trail (Capital Reef) Monument Canyon Trail (Colorado National Monument) Owachomo Bridge Trail (Natural Bridges)	13.2 11.8 9.7 11.8	
Manning et al. 1999c	Statue of Liberty/Ellis Island National Monument, NY, NJ	Visitors	• Waiting time to get into the statue • Waiting time to buy a ferry ticket • Waiting time to board the ferry	45 min. 19.3 min. 22.7 min.	
Manning et al. 1998c	Yosemite National Park, CA	Visitors	PAOT at attraction sites *Long form* • Base of Bridalveil Fall • Glacier Point *Short form* • Base of Bridalveil Fall • Glacier Point PPV on trails *Long form* • Trail to Bridalveil Fall • Trail to Mirror Lake *Short form* • Trail to Bridalveil Fall • Trail to Mirror Lake	 20 42 15 34 18 34 13 18.5	
Borrie et al. 1999	Yellowstone National Park, WY, MT, ID	Winter Visitors	Snowmobile encounters/hour	33	
Vande Kamp et al. 2003	Kenai Fjords National Park, AK	Visitors	PAOT at Exit Glacier *Long form* *Short form*	 33.0 22.8	 20.0
Manning et al. 2002c	Mesa Verde National Park, CO	Visitors	PAOT at attraction sites *Long form* • Spruce Tree House • Museum • Step House • Sun Point Overlook *Short form* • Spruce Tree House • Museum • Step House • Sun Point Overlook	 53.0 23.0 24.0 38.0 44.7 19.0 16.4 30.7	

| Study | Area | Respondents | Indicator | Normative standard | |
				Mean	Median
Manning et al. 2002c *cont'd*			PPV on Spruce Tree House Trail		
			Long form	31.0	
			Short form	26.8	
			Group tour characteristics at Balcony House		
			• Tour frequency	>15 min.	
			• Tour group size	52	
			• Tour length	>45 min.	
			Group tour characteristics at Long House		
			• Tour frequency	>15 min.	
			• Tour group size	52	
			• Tour length	<2 hrs.	
			Number of tour groups at Cliff Palace	3 groups	
Manning et al. 2002d	Acadia National Park, Schoodic Peninsula, ME	Visitors	Trail impacts	Photo 2.8	
			VPV on the scenic road	7.5	
			PAOT at attraction sites		
			• Schoodic Point	70.1	
			• Frazer Point	85.0	
Manning et al. 2002e	Blue Ridge Parkway, VA, NC	Visitors	VPV on the road	7.0	
			PAOT at scenic overlooks	34.5	
Park Studies Lab	Hawaii Volcanos National Park, HI	Visitors	PAOT at attraction sites		
			• Visitor Center	32.0	
			• Lava flow at the end of Chain of Craters Road	331.5	
			• Thurston Lava Tube	13.0	
			Cars parked on Chain of Craters Road	22.3	
			PPV on Thurston Lava Tube trail	9.3	
			Walking time from Chain of Craters Road to lava flow	24.5 min.	20.0 min.
Bacon et al. 2004	Acadia National Park, Isle au Haut, ME	Visitors	Encounters/day	5.6	
			Trail Impacts	Photo 2.7	
			Visitor-caused trails	Photo 2.1	
			Trail development	Photo 3.4	
		Residents	Encounters/day	1.9	2.0
			Trail impacts	Photo 2.6	
			Visitor-caused trails	Photo 2.3	
			Trail development	Photo 3.3	
Manning et al. 2004b	Zion National Park, UT	Visitors (2002)	Maximum group size		
			• Nonpermitted back-country users	11	
			• Permitted day users	9	
			• Overnight backcountry users	9	

				Normative standard	
Study	*Area*	*Respondents*	*Indicator*	*Mean*	*Median*
		Visitors (2003)	Number of other hikers seen/day		
			• Nonpermitted day users	73.8	
			• Nonpermitted Narrows users	21.6	
			• Weeping Rock day users	60	
			Other groups of hikers seen or heard/day		
			• Overnight users	7.5	
			• Permitted Narrows users	7.5	
			• Permitted canyon day users	6.4	
Manning and Budruk 2004	Boston Harbor Islands, MA	Visitors	Bumpkin Island		
			Hikers seen/hour	18.7	12.0
			Trail impacts		
			• Long form	Photo 3.6	
			• Short form	Photo 2.5	3.0
			Campsite impacts		
			• Long form	Photo 3.2	
			• Short form	Photo 2.3	2.0
			Litter		
			• Long form	Photo 1.4	
			• Short form	Photo 1.2	1.0
			Frequency of ferry service	45–75 min.	
			Frequency of water shuttle service	30–90 min.	
			Georges Island		
			PAOT		
			• Long form	185	
			• Short form	156	145
			Litter		
			• Long form	Photo 1.6	
			• Short form	Photo 1.2	1.0
			Graffiti		
			• Long form	Photo 2.0	
			• Short form	Photo 1.9	2.0
			Frequency of ferry service	78 min.	
			Frequency of water shuttle service	78 min.	
			Grape Island		
			Hikers seen/hour	13.5	10.0
			Trail impacts		
			• Long form	Photo 3.3	
			• Short form	Photo 2.2	2.0
			Campsite impacts		
			• Long form	Photo 3.0	
			• Short form	Photo 2.3	2.0
			Litter		
			• Long form	Photo 1.6	
			• Short form	Photo 1.1	1.0

Study	Area	Respondents	Indicator	Normative standard	
				Mean	Median
			Frequency of ferry service	>45 min.	
			Frequency of water shuttle service	>30 min.	
			Little Brewster Island		
			PAOT		
			• Long form	95	
			• Short form	72	55
			Number of visitors on tour	33.7	30.0
			Litter		
			• Long form	Photo 1.4	
			• Short form	Photo 1.2	1.0
			Lovell's Island	19	
			Hikers seen/hour	27.5	
			Campsite impacts	Photo 3.4	
			• Long form		1.0
			• Short form	Photo 1.7	
			Litter		
			• Long form	Photo 1.5	1.0
			• Short form	Photo 1.2	
			Graffiti		
			• Long form	Photo 2.2	
			• Short form	Photo 1.5	1.0
			Frequency of ferry service	<60 min.	
			Frequency of water shuttle service	<60 min.	
			Peddock's Island		
			Hikers seen/hour	26.3	18.5
			Trail impacts	18.5	
			• Long form	Photo 4.4	
			• Short form	Photo 2.4	3.0
			Campsite impacts		
			• Long form	Photo 3.6	
			• Short form	Photo 2.5	2.0
			Litter		
			• Long form	Photo 1.6	
			• Short form	Photo 1.3	1.0
			Graffiti		
			• Long form	Photo 2.2	
			• Short form	Photo 1.5	1.0
			Frequency of ferry service	30–75 min.	
			Frequency of water shuttle service	30–75 min.	
			World's End		
			Hikers seen/hour	19.2	15.0
			Trail impacts		
			• Long form	Photo 4.8	
			• Short form	Photo 3.1	3.0
			Litter		
			• Long form	Photo 1.5	
			• Short form	Photo 1.2	1.0

				Normative standard	
Study	*Area*	*Respondents*	*Indicator*	*Mean*	*Median*
Park Studies Lab 2004	Haleakala National Park, HI	Visitors	PAOT HALEAKALA VISITOR CENTER *December 2003/January 2004*		
			• Viewing area	159	
			• Visitor Center	43	
			July 2004		
			• Viewing area	152	
			• Visitor Center	41	
			RED HILL OVERLOOK *December 2003/January 2004*		
			• Viewing area	115	
			• Shelter	56	
			July 2004		
			• Viewing area	117	
			• Shelter	55	
Hall and Roggenbuck 2002	Okeefenokee National Wildlife Center, GA, FL	Visitors	Number of motorboats		
			• Semi-open question format	5.4	2.75
			• 30 point scale	8.3	5.0
			• 60 point scale	8.5	5.0
			Number of canoes		
			• Semi-open question format	23.8	10.0
			• 30 point scale	16.0	15.0
			• 60 point scale	17.8	15.0
Inglis et al. 1999	Great Barrier Reef, Australia	Snorkelers	Maximum acceptable number of snorkelers		
			• Experienced recreationists	17.0	
			• Locals	13.0	
			• Tourists	21.0	
			• Novices	11.0	
Whittaker and Shelby 2002	Hells Canyon, ID, OR	Boaters	Optimal cfs		
			• Jet boat	9,000–12,000cfs	
			• Raft	15,000–25,000cfs	

Note: The National Park Service has compiled a list of indicators that have been used in carrying capacity and related plans for a variety of parks and protected areas. This information can be found at the following website: www.usercapacity.nps.gov/is.

PAOT = people-at-one-time; PPV = persons/viewscape; VPV = vehicles/viewscape

[a] For wilderness, semiwilderness, and undeveloped recreation, respectively

[b] Range over three river segments

[c] Range over four wilderness areas

[d] Range over four wilderness areas

[e] Range over visitors using four entry points

[f] Range over two question formats

[g] Number of visitors/hundred-meter trail segment

Source: Manning, R., M. Budruk, W. Valliere, and J. Hallo. 2005. *Research to Support Visitor Management at Muir Woods National Monument and Muir Beach: Study Completion Report.* Burlington: University of Vermont, Park Studies Lab.

References

Abbe, D., and R. Manning. 2005. *Managers' Perceptions of Wilderness Day Use: Impacts and Management Actions*. Paper presented at the George Wright Society Biennial Conference, Philadelphia, PA.

Agee, J., and D. Johnson. 1987. *Ecosystem Management for Parks and Wilderness*. Seattle: University of Washington Press.

Ajzen, I., and M. Fishbein. 1980. *Understanding Attitudes and Predicting Social Behavior*. Englewood Cliffs, NJ: Prentice-Hall.

Alder, J. 1996. Effectiveness of Education and Enforcement, Cairns Section of the Great Barrier Reef Marine Park. *Environmental Management* 20:541–51.

Alldredge, R. 1973. Some Capacity Theory for Parks and Recreation Areas. *Trends* 10:20–29.

Alpert, L., and L. Herrington. 1998. An Interactive Information Kiosk for the Adirondack Park Visitor Interpretive Center, Newcomb, NY. *Proceedings of the 1997 Northeastern Recreation Research Symposium*. USDA Forest Service General Technical Report NE-241, 265–67.

Anderson, D., D. Lime, and T. Wang. 1998. *Maintaining the Quality of Park Resources and Visitor Experiences: A Handbook for Managers*. St. Paul: University of Minnesota Cooperative Park Studies Unit.

Anderson, D., and M. Manfredo. 1986. Visitor Preferences for Management Actions. *Proceedings—National Wilderness Research Conference: Current Research*. USDA Forest Service General Technical Report INT-212, 314–19.

Anderson, F., and N. Bonsor. 1974. Allocation, Congestion, and the Valuation of Recreational Resources. *Land Economics* 50:51–57.

Bacon, J., D. Laven, S. Lawson, R. Manning, and W. Vallere. 2004. *Research to Support Carrying Capacity Analysis at Isle au Haut, Acadia National Park: Study Completion Report*. Burlington: University of Vermont, Park Studies Lab.

Bacon, J., R. Manning, D. Johnson, and M. Vande Kamp. 2001. Norm Stability: A Longitudinal Analysis of Crowding and Related Norms in the Wilderness of Denali National Park and Preserve. *The George Wright Forum* 18(3):62–71.

Bamford, T., R. Manning, L. Forcier, and E. Koenemann. 1988. Differential Campsite Pricing: An Experiment. *Journal of Leisure Research* 20:324–42.

Banks, J., and Carson, J. 1984. *Discrete-Event System Simulation*. Englewood Cliffs, NJ: Prentice Hall.

Basman, C., M. Manfredo, S. Barro, J. Vaske, and A. Watson. 1996. Norm Accessibility:

An Exploratory Study of Backcountry and Frontcountry Recreational Norms. *Lesisure Sciences* 18:177–91.

Bateson, J., and M. Hui. 1992. The Ecological Validity of Photographic Slides and Video-tapes in Simulating the Service Setting. *Journal of Consumer Research* 19:271–81.

Becker, R., D. Berrier, and G. Barker. 1985. Entrance Fees and Visitation levels. *Journal of Park and Recreation Administration* 3:28–32.

Becker, R., and A. Jubenville. 1982. Forest Recreation Management. *Forest Science.* New York: John Wiley & Sons, 335–55.

Becker, R., A. Jubenville, and G. Burnett. 1984. Fact and Judgment in the Search for a Social Carrying Capacity. *Leisure Sciences* 6:475–86.

Behan, R. 1974. Police State Wilderness: A Comment on Mandatory Wilderness Permits. *Journal of Forestry* 72:98–99.

Behan, R. 1976. Rationing Wilderness Use: An Example from Grand Canyon. *Western Wildlands* 3:23–26.

Belnap, J. 1998. Choosing Indicators of Natural Resource Condition: A Case Study in Arches National Park, Utah, USA. *Environmental Management* 22:635–42.

Blumer, H. 1936. Social Attitudes and Nonsymbolic Interaction. *Journal of Educational Sociology* 9:515–23.

Borkan, R., and A. Underhill. 1989. Simulating the Effects of Glen Canyon Dam Releases on Grand Canyon River Trips. *Environmental Management* 13:347–54.

Borrie, W., W. Freimund, M. Davenport, R. Manning, W. Valliere, and B. Wang. 1999. *Winter Visit and Visitor Characteristics of Yellowstone National Park.* Missoula: University of Montana, School of Forestry.

Boteler, F. 1984. Carrying Capacity as a Framework for Managing Whitewater Use. *Journal of Park and Recreation Administration* 2:26–36.

Bowker, J., H. Cordell, and C. Johnson. 1999. User Fees for Recreation Services on Public Lands: A National Assessment. *Journal of Park and Recreation Administration* 17(3): 1–14.

Bowker, J., and V. Leeworthy. 1998. Accounting for Ethnicity in Recreation Demand: A Flexible Count Data Approach. *Journal of Leisure Research* 30(1):64–79.

Bowman, E. 1971. The Cop Image. *Parks and Recreation* 6:35–36.

Boyers, L., M. Fincher, and J. van Wagtendonk. 1999. Twenty-Eight Years of Campsite Monitoring in Yosemite National Park. *Wilderness Science in a Time of Change Proceedings.* USDA Forest Service General Technical Report RMRS-P-000.

Bright, A. 1994. Information Campaigns that Enlighten and Influence the Public. *Parks and Recreation* 29:49–54.

Bright, A., M. Fishbein, M. Manfredo, and A. Bath. 1993. Application of the Theory of Reasoned Action to the National Park Service's Controlled Burn Policy. *Journal of Leisure Research* 25:263–80.

Bright, A., and M. Manfredo. 1995. Moderating Effects of Personal Importance on the Accessibility of Attitudes toward Recreation Participation. *Leisure Sciences* 17:281–94.

Bright, A., M. Manfredo, M. Fishbein, and A. Bath. 1993. Application of the Theory of Learned Action to the National Park Service's Controlled Burn Policy. *Journal of Leisure Research* 25:263–80.

Brown, C., J. Halstead, and A. Luloff. 1992. Information as a Management Tool: An Evaluation of the Pemigewasset Wilderness Management Plan. *Environmental Management* 16:143–48.

Brown, J. 2000. Privatizing the University—the New Tragedy of the Commons. *Science*

290:1701–02.

Brown, P. 1977. Information Needs for River Recreation Planning and Management. *Proceedings: River Recreation Management and Research Symposium.* USDA Forest Service General Technical Report NC-28, 193–201.

Brown, P., and J. Hunt. 1969. The Influence of Information Signs on Visitor Distribution and Use. *Journal of Leisure Research* 1:79–83

Brown, P., S. McCool, and M. Manfredo. 1987. Evolving Concepts and Tools for Recreation User Management in Wilderness. *Proceedings—National Wilderness Research Conference: Issues, State-of-Knowledge, Future Directions.* USDA Forest Service General Technical Report INT-220, 320–46.

Brown, T., M. Richards, T. Daniel, and D. King. 1989. Recreation Participation and the Validity of Photo-Based Preference Judgments. *Journal of Leisure Research* 21:40–60.

Brunson, M., B. Shelby, and J. Goodwin. 1992. Matching Impacts with Standards in the Design of Wilderness Permit Systems. *Defining Wilderness Quality: The Role of Standards in Wilderness Management—A Workshop Proceedings.* USDA Forest Service General Technical Report PNW-305, 101–6.

Buchanan, T., J. E. Christensen, and R. J. Burdge. 1981. Social Groups and the Meanings of Outdoor Recreation Activities. *Journal of Leisure Research* 13(3):254–66.

Buckley, R. 1999. An Ecological Perspective on Carrying Capacity. *Annals of Tourism Research* 26:705–08.

Buckley, R. 2004. *Environmental Impacts of Ecotourism.* Cambridge, MA: CAB International.

Budruk, M., and R. Manning. 2003. Crowding Related Norms in Outdoor Recreation: U.S. Versus International Visitors. *Proceedings of the 2002 Northeastern Recreation Research Symposium.* USDA Forest Service General Technical Report NE-302, 216–21.

Budruk, M., and R. Manning. 2004. Indicators and Standards of Quality at an Urban-Proximate Park: Litter and Graffiti at Boston Harbor Islands National Recreation Area. *Proceedings of the 2003 Northeastern Recreation Research Symposium.* USDA Forest Service General Technical Report NE-317, 24–31.

Budruk, M., and R. Manning. 2006. Indicators and Standards of Quality at an Urban-Proximate Park: Litter and Graffiti at Boston Harbor Islands National Recreation Area. *Journal of Park and Recreation Administration* 24:1–3.

Burch Jr., W. R. 1964. Two Concepts for Guiding Recreation Decisions. *Journal of Forestry* 62:707–12.

Burch Jr., W. R. 1969. The Social Circles of Leisure: Competing Explanations. *Journal of Leisure Research* 1:125–47.

Burch Jr., W. R. 1981. The Ecology of Metaphor—Spacing Regularities for Humans and Other Primates in Urban and Wildland Habitats. *Leisure Sciences* 4:213–31.

Burch Jr., W. 1984. Much Ado About Nothing—Some Reflections on the Wider and Wilder Implications of Social Carrying Capacity. *Leisure Sciences* 6:487–96.

Burde, J., J. Peine, J. Renfro, and K. Curran. 1988. Communicating with Park Visitors: Some Successes and Failures at Great Smoky Mountains National Park. *National Association of Interpretation 1988 Research Monograph*, 7–12.

Burgess, R., R. Clark, and J. Hendee. 1971. An Experimental Analysis of Anti-litter Procedures. *Journal of Applied Behavior Analysis* 4:71–75.

Burrell, G., and F. Siebert. 1916. Gases Found in Coal Mines. *Miners Circular 14.* Bureau of Mines, U.S. Department of the Interior, Washington, DC.

Bury, R. 1976. Recreation Carrying Capacity—Hypothesis or Reality? *Parks and Recreation* 11:23–25, 56–58.

Bury, R., and C. Fish. 1980. Controlling Wilderness Recreation: What Managers Think and Do. *Journal of Soil and Water Conservation* 35:90–93.

Cable, T., D. Knudson, E. Udd, and D. Stewart. 1987. Attitude Changes as a Result of Exposure to Interpretive Messages. *Journal of Park and Recreation Administration* 5:47–60.

Campbell, F., J. Hendee, and R. Clark. 1968. Law and Order in Public Parks. *Parks and Recreation* 3:51–55.

Carmines, E., and R. Zeller. 1979. *Reliability and Validity Assessment.* Thousand Oaks, CA: Sage Publications.

Caughley, G. 1979. What Is this Thing Called Carrying Capacity? *North American Elk: Ecology, Behavior and Population.* Laramie: University of Wyoming.

Chavez, D. 1996. Mountain Biking: Direct, Indirect, and Bridge-Building Management Styles. *Journal of Park and Recreation Administration* 14:21–35.

Cheek Jr., N. H. 1971. Toward a Sociology of Not-Work. *Pacific Sociological Review* 14:245–58.

Chenoweth, R. 1990. Image Capture Technology and Aesthetic Regulation of Landscapes Adjacent to Public Lands. In *Managing America's Enduring Wilderness Resource.* St. Paul: University of Minnesota, 563–68.

Christensen, H. 1981. *Bystander Intervention and Litter Control: An Experimental Analysis of an Appeal to help Program.* USDA Forest Service Research Paper PNW-287.

Christensen, H. 1986. Vandalism and Depreciative Behavior. *A Literature Review: The President's Commission on Americans Outdoors.* Washington, DC: U.S. Government Printing Office, M-73–M-87.

Christensen, H., and R. Clark. 1983. Increasing Public Involvement to Reduce Depreciative Behavior in Recreation Settings. *Leisure Sciences* 5:359–78.

Christensen, H., and D. Dustin. 1989. Reaching Recreationists at Different Levels of Moral Development. *Journal of Parks and Recreation Administration* 7:72–80.

Christensen, H., D. Johnson, and M. Brookes. 1992. *Vandalism: Research, Prevention, and Social Policy.* USDA Forest Service General Technical Report PNW-293.

Christensen, N., A. Bartuska, J. Brown, S. Carpenter, C. D'Antonio, R. Francis, J. Franklin, J. MacMahon, R. Noss, D. Parsons, C. Peterson, M. Turner, and R. Woodmansee. 1996. Report of the Ecological Society of America Committee on the Scientific Basis for Ecosystem Management. *Ecological Applications* 6:665–91.

Christensen, N., W. Stewart, and D. King. 1993. National Forest Campgrounds: Users Willing to Pay More. *Journal of Forestry* 91:43–47.

Cialdini, R. B., C. A. Kallgren, and R. R. Reno. 1991. A Focus Theory of Normative Conduct: A Theoretical Refinement and Reevaluation of the Role of Norms in Human Behavior. *Advances in Experimental Social Psychology* 24:201–34.

Cialdini, R. B., R. R. Reno, and C. A. Kallgren. 1990. A Focus Theory of Normative Conduct: Recycling the Concept of Norms to Reduce Littering in Public Places. *Journal of Personality and Social Psychology* 58:1015–26.

Clark, R., R. Burgess, and J. Hendee. 1972a. The Development of Anti-Litter Behavior in a Forest Campground. *Journal of Applied Behavior Analysis* 5:1–5.

Clark, R., J. Hendee, and R. Burgess. 1972b. The Experimental Control of Littering. *Journal of Environmental Education* 4:22–28.

Clark, R., J. Hendee, and F. Campbell. 1971. *Depreciative Behavior in Forest Campgrounds: An Exploratory Study.* USDA Forest Service Research Paper PNW-161.

Clark, R., and G. Stankey. 1979. *The Recreation Opportunity Spectrum: A Framework for Planning, Management, and Research.* USDA Forest Service Research Paper PNW-98.

Clements, F. 1920. *Plant Indicators*. Washington, DC: Carnegie Institute.

Cockrell, D., and W. McLaughlin. 1982. Social Influences on Wild River Recreationists. *Forest and River Recreation: Research Update*. St. Paul: University of Minnesota Agricultural Experiment Station, Miscellaneous Publication 18, 140–45.

Cockrell, D., and J. Wellman. 1985a. Democracy and Leisure: Reflections on Pay-As-You-Go Outdoor Recreation. *Journal of Park and Recreation Administration* 3:1–10.

Cockrell, D., and J. Wellman. 1985b. Against the Running Tide: Democracy and Outdoor Recreation User Fees. *Proceedings of the 1985 National Outdoor Recreation Trends Symposium, Volume II*. Atlanta: U.S. National Park Service, 193–205.

Coffey, A., and P. Atkinson. 1996. *Making Sense of Qualitative Data: Complimentary Research Strategies*. Thousand Oaks, CA: Sage Publications.

Cohen, J. 1995. Population Growth and Earth's Human Carrying Capacity. *Science* 269:341–46.

Cohen, J. 1997. Population, Economics, Environment and Culture: An Introduction to Human Carrying Capacity. *Journal of Applied Ecology* 34:1325–33.

Cole, D. 1987. Research on Soil and Vegetation in Wilderness: A State-of-Knowledge Review. *Proceedings—National Wilderness Research Conference: Issues, State-of-Knowledge, Future Directions*. USDA Forest Service General Technical Report INT-200, 135–77.

Cole, D. 1989. *Wilderness Campsite Monitoring Methods: A Sourcebook*. USDA Forest Service General Technical Report INT-259.

Cole, D. 1993. Wilderness Recreation Management. *Journal of Forestry* 91:22–24.

Cole, D. 1994. *The Wilderness Threats Matrix: A Framework for Assessing Impacts*. USDA Forest Service General Technical Report INT-475.

Cole, D. 1996. *Wilderness Recreation Use Trends, 1965–1994*. USDA Forest Service Research Paper INT-488.

Cole, D. 2002. Simulation of Recreational Use in Backcountry Settings: An Aid to Management Planning. *Proceedings of the International Conference on Monitoring and Management of Visitor Flows in Recreational and Protected Areas*. Bodenkultur University, Vienna, 478–82.

Cole, D. 2005. *Computer Simulation Modeling of Recreation Use: Current Status, Case Studies, and Future Directions*. USDA Forest Service General Technical Report RMRS-GTR-143.

Cole, D., K. Cahill, and M. Hof. 2005. Why Model Recreation Use? *Computer Simulation Modeling of Recreation Use Current Status, Case Studies, and Future Directions*. USDA Forest Service General Technical Report RMRS-GTR-143,1–2.

Cole, D., T. Hammond, and S. McCool. 1997a. Information Quality and Communication Effectiveness: Low-Impact Messages on Wilderness Trailhead Bulletin Boards. *Leisure Sciences* 19:59–72.

Cole, D., M. Peterson, and R. Lucas. 1987. *Managing Wilderness Recreation Use*. USDA Forest Service General Technical Report INT-230.

Cole, D., and B. Rang. 1983. Temporary Campsite Closure in the Selway-Bitterroot Wilderness. *Journal of Forestry* 81:729–32.

Cole, D., A. Watson, T. Hall, and D. Spildie. 1997b. *High-Use Destinations in Wilderness: Social and Bio-Physical Impacts, Visitor Responses, and Management Options*. USDA Forest Service Research Paper INT-496.

Cole, D., A. Watson, and J. Roggenbuck. 1995. *Trends in Wilderness Visitors and Visits: Boundary Waters Canoe Area, Shining Rock, and Desolation Wilderness*. USDA Forest Service Research Paper INT-483.

Commission on Sustainable Development. 2001. *Indicators of Sustainable Development: Guidelines and Methodologies.* New York: United Nations Division for Sustainable Development.

Confer, J., J. Absher, A. Graefe, and A. Hille. 1999. Relationships Between Visitor Knowledge of "Leave No Trace" Minimum Impact Practices and Attitudes Toward Selected Management Actions. *Proceedings of the 1998 Northeastern Recreation Research Symposium.* USDA Forest Service General Technical Report NE 255, 142–46.

Confer, J., A. Mowen, A. Graefe, and J. Absher. 2000. Magazines as Wilderness Information Sources: Assessing User's General Wilderness Knowledge and Specific Leave No Trace Knowledge. *Proceedings from the National Wilderness Science Conference, A Time of Change.* U.S. Department of Agriculture, Forest Service, Rocky Mountain Research Station General Technical Report P-15,(4): 193–97.

Connors, E. 1976. Public Safety in Park and Recreation Settings. *Parks and Recreation* 11:20–21, 55–56.

Crompton, J., and C. Lue. 1992. Patterns of Equity Preferences Among Californians for Allocating Park and Recreation Resources. *Leisure Sciences* 14:227–46.

Crompton, J., and B. Wicks. 1988. Implementing a Preferred Equity Model for the Delivery of Leisure Services in the U.S. Context. *Leisure Studies* 7: 287–304.

Daily, G., and P. Ehrlich. 1992. Population, Sustainability, and Earth's Carrying Capacity. *BioScience* 42:761–71.

Daniel, T., and R. Boster. 1976. *Measuring Landscape Esthetics: The Scenic Beauty Estimation Method.* Research Paper RM-167. Fort Collins, CO: USDA Forest Service, Rocky Mountain Forest and Range Experiment Station.

Daniel, T., T. Brown, D. King, M. Richards, and W. Stewart. 1989. Perceived Scenic Beauty and Contingent Valuation of Forest Campgrounds. *Forest Science* 35:76–90.

Daniel, T., and R. Gimblett. 2000. Autonomous Agents in the Park: An Introduction to the Grand Canyon River Trip Simulation Model. *International Journal of Wilderness* 6:39–43.

Daniel, T., and W. Ittleson. 1981. Conditions for Environmental Perception Research: Comment on "the Psychological Representation of Molar Physical Environments" by Ward and Russell. *Journal of Experimental Psychology* 110:153–57.

Daniel, T., and M. Meitner. 2001. Representational Validity of Landscape Visualizations: The Effects of Graphical Realisim on Perceived Scenic Beauty of Forest Vistas. *Journal of Environmental Psychology* 21:61–72.

Daniels, S. 1987. Marginal Cost Pricing and the Efficient Provision of Public Recreation. *Journal of Leisure Research* 19:22–34.

Dasman, R. 1964. *Wildlife Biology.* New York: John Wiley & Sons.

Davidson, C. 2000. Economic Growth and the Environment: Alternatives to the Limits Paradigm. *BioScience* 50:433–44.

Deitz, T. 2005. The Darwinian Trope in the Drama of the Commons: Variations on Some Themes by the Ostroms. *Journal of Economic Behavior and Organization* 57:205–25.

Dennis, D. 1998. Analyzing Public Inputs to Multiple Objective Decisions on National Forests Using Conjoint Analysis. *Forest Science* 44:421–29.

Desvousges, W. H., V. K. Smith, and M. P. McGivney. 1983. A Comparison of Alternative Approaches for Estimating Recreation and Related Benefits of Water Quality Improvements. U.S. Environmental Protection Agency, EPA 230-05-83-001.

Dhondt, A. 1988. Carrying Capacity: A Confusing Concept. *Acta Ecologica* 9:337–46.

Donnelly, M. P., J. J. Vaske, D. Whittaker, and B. Shelby. 2000. Toward an Understanding of Norm Prevalence: A Comparative Analysis of 20 Years of Research. *Environmental Management* 25:403–14.

Dottavio, F. D., J. O'Leary, and B. Koth. 1980. The Social Group Variable in Recreation Participation Studies. *Journal of Leisure Research* 12:357–67.

Doucette, J., and D. Cole. 1993. *Wilderness Visitor Education: Information About Alternative Techniques.* USDA Forest Service General Technical Report INT-295.

Doucette, J., and K. Kimball. 1990. Passive Trail Management in Northeastern Alpine Zones: A Case Study. *Proceedings of the 1990 Northeastern Recreation Research Symposium.* USDA Forest Service General Technical Report NE-145, 195–201.

Dowell, D., and S. McCool. 1986. Evaluation of a Wilderness Information Dissemination Program. *Proceedings—National Wilderness Research Conference: Current Research.* USDA Forest Service General Technical Report INT-212, 494–500.

Driver, B. L. 1984. Public Responses to User Fees at Public Recreation Areas. *Proceedings: Fees for Outdoor Recreation on Lands Open to the Public.* Gorham, NH: Appalachian Mountain Club, 47–51.

Driver, B. L., and J. Bassett. 1975. Defining Conflicts Among River Users: A Case Study of Michigan's Au Sable River. *Naturalist* 26:19–23.

Duncan, G., and S. Martin. 2002. Comparing the Effectiveness of Interpretive and Sanction Messages for Influencing Wilderness Visitors' Intended Behavior. *International Journal of Wilderness* 8:20–25. Dustin, D. 1986. Outdoor Recreation: A Question of Equity. *Forum for Applied Research and Public Policy* 1:62–67.

Dustin, D., and R. Knopf. 1989. Equity Issues in Outdoor Recreation. *Outdoor Recreation Benchmark 1988: Proceedings of the National Outdoor Recreation Forum.* USDA Forest Service General Technical Report SE-52, 467–71.

Dustin, D., and L. McAvoy. 1980. "Hardening" National Parks. *Environmental Ethics* 2:29–44.

Dustin, D., and L. McAvoy. 1984. The Limitation of the Traffic Light. *Journal of Park and Recreation Administration* 2:8–32.

Dustin, D., L. McAvoy, and L. Beck. 1986. Promoting Recreationist Self-Sufficiency. *Journal of Park and Recreation Administration* 4:43–52.

Dustin, D., L. McAvoy, and J. Schultz. 1987. Beware of the Merchant Mentality. *Trends* 24:44–46.

Echelberger, H., R. Leonard, and S. Adler. 1983. Designated-Dispersed Tentsites. *Journal of Forestry* 81:90–91, 105.

Echelberger, H., R. Leonard, and M. Hamblin. 1978. *The Trail Guide System as a Backcountry Management Tool.* USDA Forest Service Research Note NE-266.

Edwards, R., and C. Fowle. 1955. The Concept of Carrying Capacity. *Transactions of the 20th North American Wildlife Conference.* Washington, DC: Wildlife Management Institute, 589–602.

Ehrlich, P. 1968. *The Population Bomb.* New York: Ballantine Books.

Ehrlich, P., and J. Holdren. 1971. Impact of Population Growth. *Science* 171:1212–17.

Emmett, J., M. Havitz, and R. McCarvill. 1996. A Price Subsidy Policy for Socio-Economically Disadvantaged Recreation Participants. *Journal of Park and Recreation Administration* 14:63–80.

Environment Canada and Park Service. 1991. *Selected Readings on the Visitor Activity Management Process.* Ottawa, Ontario: Environment Canada.

Ewert, A., and D. Hood. 1995. Urban-Proximate and Urban-Distant Wilderness: An Exploratory Comparison Between Two "Types" of Wilderness. *Journal of Park and Recreation Administration* 13:73–85.

Farley, J., and H. Daly. 2004. *Ecological Economics: Principles and Applications.* Washington, DC: Island Press.

Farrell, T., and J. Marion. 1998. *An Evaluation of Camping Impacts and their Management at Isle Royale National Park.* United States Department of Interior, National Park Service Research/Resources Management Report.

Fazio, J. 1979a. Communication with the Wilderness User. *Wildlife and Range Science Bulletin Number 28*, Moscow, ID: University of Idaho, College of Forestry.

Fazio, J. 1979b. Agency Literature as an Aid to Wilderness Management. *Journal of Forestry* 77:97–98.

Fazio, J., and D. Gilbert. 1974. Mandatory Wilderness Permits: Some Indications of Success. *Journal of Forestry* 72:753–56.

Fazio, J., and Ratcliffe, R. 1989. Direct-Mail Literature as a Method to Reduce Problems of Wild River Management. *Journal of Park and Recreation Administration* 7:1–9.

Fedler, A., and A. Miles. 1989. Paying for Backcountry Recreation: Understanding the Acceptability of Use Fees. *Journal of Park and Recreation Administration* 7:35–46.

Feeny, D., F. Berkes, B. McCay, and J. Acheson. 1990. The Tragedy of the Commons: Twenty-Two Years Later. *Human Ecology* 18:1–19.

Field, D. R., and J. O'Leary. 1973. Social Groups as a Basis for Assessing Participation in Selected Water Activities. *Journal of Leisure Research* 5:16–25.

Fish, C., and R. Bury. 1981. Wilderness Visitor Management: Diversity and Agency Policies. *Journal of Forestry* 79:608–12.

Fishbein, M., and I. Ajzen. 1975. *Belief, Attitude, Intention and Behavior: An Introduction to Theory and Research.* Reading, MA: Addison-Wesley.

Flekke, G., L. McAvoy, and D. Anderson. 1996. The Potential of an Expert System to Address Congestion and Crowding in the National Park System. *Crowding and Congestion in the National Park System: Guidelines for Research and Management.* St. Paul: University of Minnesota Agricultural Experiment Station Publication 86-1996, 132–41.

Force, J., and G. Machlis. 1997. The Human Ecosystem, Part II: Social Indicators in Ecosystem Management. *Society and Natural Resources* 10:360–82.

Fractor, D. 1982. Evaluating Alternative Methods for Rationing Wilderness Use. *Journal of Leisure Research* 14:341–49.

Freimund, W., S. Peel, R. Manning, and J. Bradybaugh. 2004. The Wilderness Experience as Purported by Planning Compared to Visitors at Zion National Park. *Protecting Our Diverse Heritage: The Role of Parks, Protected Areas, and Cultural Sites.* Proceedings of the 2003 George Wright Society/National Park Service Joint Conference. Hancock, MI: The George Wright Society.

Freimund, W., J. Vaski, M. Donnelly, and T. Miller. 2002. Using Video Surveys to Access Dispersed Backcountry Visitors' Norms. *Leisure Sciences* 24:349–62.

Frissell, S., and D. Duncan. 1965. Campsite Preference and Deterioration. *Journal of Forestry* 63:256–60.

Frissell, S., and G. Stankey. 1972. Wilderness Environmental Quality: Search for Social and Ecological Harmony. *Proceedings of the Society of American Foresters Annual Conference.* Hot Springs, AR: Society of American Foresters, 170–83.

Frost, J., and S. McCool. 1988. Can Visitor Regulation Enhance Recreational Experiences? *Environmental Management* 12:5–9.

Gibbs, K. 1977. Economics and Administrative Regulations of Outdoor Recreation Use. *Outdoor Recreation: Advances in Application of Economics.* USDA Forest Service General Technical Report WO-2, 98–104.

Gilbert, G., G. Peterson, and D. Lime. 1972. Towards a Model of Travel Behavior in the Boundary Waters Canoe Area. *Environment and Behavior* 4:131–57.

Gilligan, C. 1982. *In a Different Voice*. Cambridge, MA: Harvard University Press.

Gimblett, R., M. Richards, and R. Itami. 2000. RBSim: Geographic Simulation of Wilderness Recreation Behavior. *Journal of Forestry* 99:36–42.

Glass, R., and T. More. 1992. *Satisfaction, Valuation, and Views Toward Allocation of Vermont Goose Hunting Opportunities*. USDA Forest Service Research Paper NE-668.

Glick, T. 1970. *Irrigation and Society in Medieval Valencia*. Cambridge, MA: Harvard University Press.

Godin, V., and R. Leonard. 1977a. *Permit Compliance in Eastern Wilderness: Preliminary Results*. USDA Forest Service Research Note NE-238.

Godin, V., and R. Leonard. 1977b. Design Capacity for Backcountry Recreation Management Planning. *Journal of Soil and Water Conservation* 32:161–64.

Godin, V., and R. Leonard. 1979. Management Problems in Designated Wilderness Areas. *Journal of Soil and Water Conservation* 34:141–43.

Gordon, J. 1954. The Economic Theory of a Common-Property Resource: The Fishery. *Journal of Political Economy* 62:124–42.

Graefe, A., F. Kuss, and J. Vaske. 1990. *Visitor Impact Management: The Planning Framework*. Washington, DC: National Parks and Conservation Association.

Graefe, A., J. Vaske, and F. Kuss. 1984. Social Carrying Capacity: An Integration and Synthesis of Twenty Years of Research. *Leisure Sciences* 8:275–95.

Gramann, J., and G. Vander Stoep. 1987. Prosocial Behavior Theory and Natural Resource Protection: A Conceptual Synthesis. *Journal of Environmental Management* 24:247–57.

Grasmick, H., B. Blackwell, T. Barsik, and S. Mitchell. 1993. Changes in Perceived Threats of Shame, Embarrassment, and Legal Sanctions for Interpersonal Violence. *Violence and Victims* 8:313–25.

Greco, G., and L. Floridi. 2004. The Tragedy of the Digital Commons. *Ethics and Information Technology* 6:73–81.

Green, P., and V. Srinivasan. 1978. Conjoint Analysis in Consumer Research: Issues and Outlook. *Journal of Consumer Research* 5:103–23.

Griest, D. 1975. Risk Zone Management: A Recreation Area Management System and Method of Measuring Carrying Capacity. *Journal of Forestry* 73:711–14.

Grumbine, R. 1994. What Is Ecosystem Management? *Conservation Biology* 8:27–38.

Haas, G., B. Driver, P. Brown, and R. Lucas. 1987. Wilderness Management Zoning. *Journal of Forestry* 85:17–22.

Hadley, L. 1971. Perspectives on Law Enforcement in Recreation Areas. *Recreation Symposium Proceedings*. Upper Darby, PA: USDA Forest Service Northeastern Forest Experiment Station, 156–60.

Hadwen, S., and L. Palmer. 1922. *Reindeer in Alaska*. U.S. Department of Agriculture Bulletin 1089.

Hall, T., and J. Roggenbuck. 2002. Response Format Effects in Questions about Norms: Implications for the Reliability and Validity of the Normative Approach. *Leisure Sciences* 24:325–38.

Hall, T., and B. Shelby. 1996. Who Cares About Encounters? Differences Between Those With and Without Norms. *Leisure Sciences* 18:7–22.

Hall, T., B. Shelby, and D. Rolloff. 1996. Effect of Varied Question Format on Boaters' Norms. *Leisure Sciences* 18:193–204.

Hallo, J., R. Manning, and W. Valliere. 2005. Acadia National Park Scenic Roads: Estimating the Relationship Between Increasing Use and Potential Standards of Quality. *Computer Simulation Modeling of Recreation Use: Current Status, Case Studies, and Future Directions*. USDA Forest Service General Technical Report RMRS-GTR-143, 55–57.

Hammitt, W. E. 1982. Cognitive Dimensions of Wilderness Solitude. *Environment and Behavior* 14:478–93.

Hammitt, W., and D. Cole. 1998. *Wildland Recreation: Ecology and Management*. New York: John Wiley & Sons.

Hammitt, W., and M. Patterson. 1991. Coping Behavior to Avoid Visual Encounters: Its Relationship to Wildland Privacy. *Journal of Leisure Research* 23:225–37.

Hammitt, W., and W. Rutlin. 1995. Use Encounter Standards and Curves for Achieved Privacy in Wilderness. *Leisure Sciences* 17:245–62.

Hardin, G. 1968. The Tragedy of the Commons. *Science* 162:1243–48.

Hardin, G. 1986. Cultural Carrying Capacity: A Biological Approach to Human Problems. *BioScience* 36:599–604.

Hardin, G., and J. Baden. 1977. *Managing the Commons*. San Francisco: W.H. Freeman & Co.

Harmon, D. 1992. Using an Interactive Computer Program to Communicate with the Wilderness Visitor. *Proceedings of the Symposium on Social Aspects and Recreation Research*. USDA Forest Service General Technical Report PSW-132, 60.

Harmon, L. 1979. How to Make Park Law Enforcement Work for You. *Parks and Recreation* 14:20–21.

Harris, C., and B. Driver. 1987. Recreation User Fees: Pros and Cons. *Journal of Forestry* 85:25–29.

Harwell, R. 1987. A "No-Rescue" Wilderness Experience: What Are the Implications? *Parks and Recreation* 22:34–37.

Heberlein, T., G. Alfano, and L. Ervin. 1986. Using a Social Carrying Capacity Model to Estimate the Effects of Marina Development at the Apostle Islands National Lakeshore. *Leisure Sciences* 8:257–74.

Heinrichs, J. 1982. Cops in the Woods. *Journal of Forestry* 11:722–25, 748.

Hendee, J., R. Clark, and T. Daily. 1977. *Fishing and Other Recreation Behavior at High Mountain Lakes in Washington State*. USDA Forest Service Research Note PNW-304.

Hendee, J., and C. Dawson. 2002. *Wilderness Management*. 3rd ed. Golden, CO: Fulcrum Publishing.

Hendee, J., and R. Lucas. 1973. Mandatory Wilderness Permits: A Necessary Management Tool. *Journal of Forestry* 71:206–09.

Hendee, J., and R. Lucas. 1974. Police State Wilderness: A Comment on a Comment. *Journal of Forestry* 72:100–101.

Hendricks, B., E. Ruddell, and C. Bullis. 1993. Direct and Indirect Park and Recreation Resource Management Decision Making: A Conceptual Approach. *Journal of Park and Recreation Administration* 11:28–39.

Hershberger, R., and R. Cass. 1974. Predicting User Responses to Building. *Man-Environment Interactions: Evaluations and Applications*. Stroudsbury, PA: Dowden, Hutchinson & Ross, 117–34.

Hess, C. 2004. The Comprehensive Bibliography of the Commons. www.indiana.edu/~iascp/lforms/searchcpf.html.

Heywood, J. 1985. Large Recreation Group and Party Size Limits. *Journal of Park and Recreation Administration* 3:36–44.

Heywood, J. 1993. Behavioral Conventions in Higher Density, Day Use Wildland/Urban Recreation Settings: A Preliminary Case Study. *Journal of Leisure Research* 25:39–52.

Heywood, J. 1996. Social Regularities in Outdoor Recreation. *Leisure Sciences* 18:23–37.

Heywood, J. 2002. The Cognitive and Emotional Components of Behavior Norms in Outdoor Recreation. *Leisure Sciences* 24:271–81.

Heywood, J., and R. Engelke. 1995. Differences in Behavioral Convention: A Comparison of United States-Born and Mexican-Born Hispanics, and Anglo-Americans. *Proceedings of the Second Symposium on Social Aspects and Recreation Research*. USDA Forest Service General Technical Report PSW-156, 35–40.

Heywood, J., and W. Murdock. 2002. Social Norms in Outdoor Recreation: Searching for the Behavior-Condition Link. *Leisure Sciences* 24:283–96.

Hof, M., J. Hammett, M. Rees, J. Belnap, N. Poe, D. Lime, and R. Manning. 1994. Getting a Handle on Carrying Capacity: A Pilot Project at Arches National Park. *Park Science* 14:11–13.

Hollenhorst, S., S. Brock, W. Freimund, and M. Twery. 1993. Predicting the Effects of Gypsy Moth on Near-View Aesthetic Preferences and Recreation Behavior Intentions. *Forest Science* 39:28–40.

Hollenhorst, S., and L. Gardner. 1994. The Indicator Performance Estimator Approach to Determining Acceptable Wilderness Conditions. *Environmental Management* 18:901–06.

Hollenhorst, S., D. Olson, and R. Fortney. 1992. Use of Importance-Performance Analysis to Evaluate State Park Cabins: The Case of the West Virginia State Park System. *Journal of Park and Recreation Administration* 10:1–11.

Hollenhorst, S., and L. Stull-Gardner. 1992. The Indicator Performance Estimate (IPE) Approach to Defining Acceptable Conditions in Wilderness. *Proceedings of the Symposium on Social Aspects and Recreation Research*. USDA Forest Service General Technical Report PSW-132, 48–49.

Holling, C. 1978. *Adaptive Environmental Assessment and Management*. London: John Wiley & Sons.

Hope, J. 1971. Hassles in the Park. *Natural History* 80:20–23, 82–91.

Horsley, A. 1988. The Unintended Effects of Posted Sign on Littering Attitudes and Stated Intentions. *Journal of Environmental Education* 19:10–14.

Hosmer, D., and S. Lemeshow. 2000. *Applied Logistic Regression*. New York: John Wiley & Sons.

Huffman, M., and D. Williams. 1987. The Use of Microcomputers for Park Trail Information Dissemination. *Journal of Park and Recreation Administration* 5:35–46.

Hulbert, J., and J. Higgins. 1977. BWCA Visitor Distribution System. *Journal of Forestry* 75:338–40.

Hull, B., and W. Stewart. 1992. Validity of Photo-Based Scenic Beauty Judgments. *Journal of Environmental Psychology* 12:101–14.

Hultsman, W. 1988. Applications of a Touch-Sensitive Computer in Park Settings: Activity Alternatives and Visitor Information. *Journal of Parks and Recreation Administration* 6:1–11.

Hultsman, W., and J. Hultsman. 1989. Attitudes and Behaviors Regarding Visitor-Control Measures in Fragile Environments: Implications for Recreation Management. *Journal of Park and Recreation Administration* 7:60–69.

Inglis, G., V. Johnson, and F. Ponte. 1999. Crowding Norms in Marine Settings: A Case

Study of Snorkeling on the Great Barrier Reef. *Environmental Management* 24(3): 369–81.

Jackson, J. M. 1965. Structural Characteristics of Norms. *Current Studies in Social Psychology*, I. D. Steiner and M. F. Fishbein, eds. New York: Holt, Rinehart & Winston, 301–309.

Jacobi, C., and R. Manning. 1999. Crowding and Conflict on the Carriage Roads of Acadia National Park: An Application of the Visitor Experience and Resource Protection Framework. *Park Science* 19:22–26.

Jacobi, C., R. Manning, W. Valliere, and C. Negra. 1996. Visitor Use and Conflict on the Carriage Roads of Acadia National Park. *Proceedings of the 1995 Northeastern Recreation Research Symposium*. USDA Forest Service General Technical Report NE-218, 109–12.

Johnson, D., and M. Vande Kamp. 1996. Extent and Control of Resource Damage Due to Noncompliant Visitor Behavior: A Case Study from the U.S. National Parks. *Natural Areas Journal* 16:134–41.

Jones, P., and L. McAvoy. 1988. An Evaluation of a Wilderness User Education Program: A Cognitive and Behavioral Analysis. *National Association of Interpretation 1988 Research Monograph*, 13–20.

Jubenville, A., and R. Becker. 1983. Outdoor Recreation Management Planning: Contemporary Schools of Thought. *Recreation Planning and Management*. State College, PA: Venture Publishing, 303–19.

Keep America Beautiful. 2000. *Litter Index*. Stamford, CT: Keep America Beautiful.

Kellomaki, S., and R. Savolianen. 1984. The Scenic Value of Forest Landscape as Assessed in the Field and the Laboratory. *Landscape Planning* 11:97–108.

Kennedy, D. 2003. Sustainability and the Commons. *Science* 302:1861.

Kernan, A., and E. Drogin. 1995. The Effect of a Verbal Interpretive Message on Day User Impacts at Mount Rainier National Park. *Proceedings of the 1994 Northeastern Recreation Research Symposium*. USDA Forest Service General Technical Report NE-198, 127–29.

Kerr, G., and M. Manfredo. 1991. An Attitudinal-Based Model of Pricing for Recreation Services. *Journal of Leisure Research* 23:37–50.

Kim, S., and B. Shelby. 1998. Norms for Behavior and Conditions in Two National Park Campgrounds in Korea. *Environmental Management* 22:277–85.

Kneeshaw, K., J. Vaske, A. Bright, and J. Absher. 2004. Situational Influences of Acceptable Wildland Fire Management Actions. *Society and Natural Resources* 17:477–89.

Kohlberg, L. 1976. Moral Stages and Moral Development. *Moral Development and Behavior: Theory, Research, and Social Issues*. New York: Holt, Rinehart & Winston.

Krannich, R., B. Eisenhauer, D. Field, C. Pratt, and A. Luloff. 1999. Implications of the National Park Service Recreation Fee Demonstration Program for Park Operations and Management: Perceptions of NPS Managers. *Journal of Park and Recreation Administration* 17:35–52.

Krumpe, E., and P. Brown. 1982. Using Information to Disperse Wilderness Hikers. *Journal of Forestry* 80:360–62.

Kuhn, S. 2003. Prisoner's Dilemma. *The Stanford Encyclopedia of Philosophy*. Fall 2003 edition. http://plato.stanford.edu/archives/fall2003/entries/prisoner-dilemma/.

Kuss, F., A. Graefe, and J. Vaske. 1990. *Visitor Impact Management: A Review of Research*. Washington, DC: National Parks and Conservation Association.

Kuss, F., and J. Morgan III. 1984. Using the USLE to Estimate the Physical Carrying Capacity of Natural Areas for Outdoor Recreation Planning. *Journal of Soil and Water Conservation* 39:383–87.

Kuss, F., and J. Morgan III. 1986. A First Alternative for Estimating the Physical Carrying Capacities of Natural Areas for Recreation. *Environmental Management* 10:255–62.

LaChapelle, P., and S. McCool. 2005. Exploring the Concept of "Ownership" in Natural Resource Planning. *Society and Natural Resources* 18:279–85.

Lahart, D., and J. Barley. 1975. Reducing Children's Littering on a Nature Trail. *Journal of Environmental Education* 7:37–45.

Lake Champlain Basin Program. 2003. *Opportunities for Action: An Evolving Plan for the Future of the Lake Champlain Basin.* Grand Isle, Vermont: Lake Champlain Basin Program.

LaPage, W., P. Cormier, G. Hamilton, and A. Cormier. 1975. *Differential Campsite Pricing and Campground Attendance.* USDA Forest Service Research Paper NE-330.

Laven, D., R. Manning, and D. Krymkowski. 2005. The Relationship Between Visitor-Based Standards of Quality and Existing Conditions in Parks and Outdoor Recreation. *Leisure Sciences* 27:157–73

Law, A., and W. Kelton. 1991. *Simulation Modeling and Analysis.* New York: McGraw-Hill.

Lawson, S., A. Keily, and R. Manning. 2003a. Computer Simulation as a Tool for Developing Alternatives for Managing Crowding at Wilderness Campsites on Isle Royale. *The George Wright Forum* 20:72–82.

Lawson, S., and R. Manning. 2000. Crowding Versus Access at Delicate Arch, Arches National Park: An Indifference Curve Analysis. *Proceedings of the Third Symposium on Social Aspects and Recreation Research.* Tempe, AZ: Arizona State University,135–43.

Lawson, S., and R. Manning. 2001a. Solitude Versus Access: A Study of Tradeoffs in Outdoor Recreation Using Indifference Curve Analysis. *Leisure Sciences* 23:179–91.

Lawson, S., and R. Manning. 2001b. Evaluating Multiple Dimensions of Visitors' Tradeoffs Between Access and Crowding at Arches National Park Using Indifference Curve Analysis. *Proceedings of the 2000 Northeastern Recreation Research Symposium.* USDA Forest Service General Technical Report NE-270, 167–75.

Lawson, S., and R. Manning. 2001c. Crossing Experiential Boundaries: Visitor Preferences Regarding Tradeoffs Among Social, Resource, and Managerial Attributes of the Denali Wilderness. *The George Wright Forum* 18:10–27.

Lawson, S., and R. Manning. 2002a. Using Simulation Modeling to Facilitate Proactive Monitoring and Adaptive Management of Social Carrying Capacity at Arches National Park, Utah, USA. *Monitoring and Management of Visitor Flows in Recreational and Protected Areas.* Vienna, Austria: Bodenkultur University, 205–10.

Lawson, S., and R. Manning. 2002b. Tradeoffs Among Social, Resource, and Managerial Attributes of the Denali Wilderness Experience: A Contextual Approach to Normative Research. *Leisure Sciences* 24:297–312.

Lawson, S., and R. Manning. 2002c. Integrating Multiple Wilderness Values into a Decision-Making Model for Denali National Park and Preserve. *Monitoring and Management of Visitor Flows in Recreational and Protected Areas.* Vienna, Austria: Bodenkultur University, 136–42.

Lawson, S., and R. Manning. 2002d. Balancing Tradeoffs in the Denali Wilderness: An Expanded Approach to Normative Research Using Stated Choice Analysis. *Proceedings of the 2001 Northeastern Recreation Research Symposium.* USDA Forest Service General Technical Report NE-289, 15–24.

Lawson, S., and R. Manning. 2003a. Research to Guide Management of Wilderness Camping at Isle Royale National Park: Part I—Descriptive Research. *Journal of Park and Recreation Administration* 21:22–42.

Lawson, S., and R. Manning. 2003b. Research to Guide Management of Wilderness Camping at Isle Royale National Park: Part II—Prescriptive Research. *Journal of Park and Recreation Administration* 21:43–56

Lawson, S., R. Manning, W. Valliere, and B. Wang. 2003. Proactive Monitoring and Adaptive Management of Social Carrying Capacity in Arches National Park: An Application of Computer Simulation Modeling. *Journal of Environmental Management* 68:305–13.

Lee, K. 1993. *Compass and Gyroscope: Integrating Science and Politics for the Environment.* Washington, DC: Island Press.

Lee, R. 1972. The Social Definition of Outdoor Recreation Places. *Social Behavior, Natural Resources, and the Environment.* New York: Harper & Row, 68–84.

Lee, R. 1975. *The Management of Human Components in the Yosemite National Park Ecosystem: Final Research Report.* Berkeley: University of California.

Lee, R. 1977. Alone with Others: The Paradox of Privacy in Wilderness. *Leisure Sciences* 1:3–9.

Leopold, A. 1933. *Game Management.* New York: Charles Scribner's Sons.

Leopold, L., F. Clarke, B. Hanshaw, and J. Balsley. 1971. *A Procedure for Evaluating Environmental Impact.* Geological Survey Circular 645. Washington, DC: U.S. Geological Survey.

Leung, Y., and J. Marion. 1999a. Assessing Trail Conditions in Protected Areas: Application of a Problem Assessment Method in Great Smoky Mountains National Park, U.S.A. *Environmental Conservation* 26:270–79.

Leung, Y., and J. Marion. 1999b. The Influence of Sampling Interval on the Accuracy of Trail Impact Assessment. *Landscape and Urban Planning* 43:167–79.

Leung, Y., and J. Marion. 1999c. Characterizing Backcountry Camping Impacts in Great Smoky Mountains National Park, USA. *Journal of Environmental Management* 57:193–203.

Leung, Y., and J. Marion. 2000. Recreation Impacts and Management in Wilderness: A State-of-Knowledge Review. *Proceedings: Wilderness Science in a Time of Change.* USDA Forest Services Proceedings RMRS-P-15-Vol-5.

Leuschner, W., P. Cook, J. Roggenbuck, and R. Oderwald. 1987. A Comparative Analysis for Wilderness User Fee Policy. *Journal of Leisure Research* 19:101–14.

Lewis, H. 1980. Irrigation Societies in the Northern Philippines. *Irrigation and Agricultural Development in Asia: Perspectives from the Social Sciences.* Ithaca, NY: Cornell University Press.

Lewis, M., D. Lime, and P. Anderson. 1996a. Use of Visitor Encounter Norms in Natural Area Management. *Natural Areas Journal* 16:128–33.

Lewis, M., D. Lime, and P. Anderson. 1996b. Paddle Canoeists Encounter Norms in Minnesota's Boundary Waters Canoe Area Wilderness. *Leisure Sciences* 18:143–60.

Lewis, R. 2004. Academic Emergency Medicine and the "Tragedy of the Commons." *Academic Emerging Medicine* 11:423–27.

Lime, D. 1972. *Large Groups in the Boundary Waters Canoe Area—Their Numbers, Characteristics, and Impact.* USDA Forest Service Research Note NC-142.

Lime, D. 1977a. Principles of Recreation Carrying Capacity. *Proceedings of the Southern States Recreation Research Applications Workshop.* Asheville, NC, 122–34.

Lime, D. 1977b. When the Wilderness Gets Crowded . . . ? *Naturalist* 28:1–7.

Lime, D. 1977c. Alternative Strategies for Visitor Management of Western Whitewater

River Recreation. *Managing Colorado River Whitewater: The Carrying Capacity Strategy.* Logan: Utah State University, 146–55.

Lime, D. 1979. Carrying Capacity. *Trends* 16:37–40.

Lime, D. 1990. Image Capture Technology: An Exciting New Tool for Wilderness Managers. *Managing America's Enduring Wilderness Resource.* St. Paul: University of Minnesota, 549–52.

Lime, D. 1995. Principles of Carrying Capacity for Parks and Outdoor Recreation Areas. *Acta Envioronmentalica Universitatis Comemiane* 4:19–24.

Lime, D., D. Anderson, and J. Thompson. 2004. *Identifying and Monitoring Indicators of Visitor Experience and Resource Quality: A Handbook for Recreation Resource Managers.* St. Paul: University of Minnesota, Department of Forest Resources.

Lime, D., and G. Lorence. 1974. *Improving Estimates of Wilderness Use from Mandatory Travel Permits.* USDA Forest Service Research Paper NC-101.

Lime, D., and R. Lucas. 1977. Good Information Improves the Wilderness Experience. *Naturalist* 28:18–20.

Lime, D., R. Manning, and W. Freimund. 2001. *Finishing the Social Science Research Agenda at Arches National Park: Study Completion Report.* Burlington: University of Vermont, Park Studies Lab.

Lime, D., and G. Stankey. 1971. Carrying Capacity: Maintaining Outdoor Recreation Quality. *Recreation Symposium Proceedings.* USDA Forest Service, 174–84.

Lindberg, K., and B. Aylward. 1999. Price Responsiveness in the Developing Country Nature Tourism Context: Review and Costa Rican Case Study. *Journal of Leisure Research* 31:281–99.

Lindberg, K., and S. McCool. 1998. A Critique of Environmental Carrying Capacity as a Means of Managing the Effects of Tourism Development. *Environmental Conservation* 25:291–92.

Lindberg, K., S. McCool, and G. Stankey. 1997. Rethinking Carrying Capacity. *Annals of Tourism Research.* 24:401–65.

Lloyd, W. 1833. *Two Lectures on the Checks to Population.* Oxford, England: Oxford University Press.

Louviere, J., D. Hensher, and J. Swait. 2000. *Stated Choice Methods: Analysis and Application.* Cambridge, U.K.: Cambridge University Press.

Louviere, J., and H. Timmermans. 1990. Stated Preference and Choice Models Applied to Recreation Research: A Review. *Leisure Sciences* 12:9–32.

Lucas, R. 1964. *The Recreational Capacity of the Quetico-Superior Area.* USDA Forest Service Research Paper LS-15.

Lucas, R. 1980. *Use Patterns and Visitor Characteristics, Attitudes, and Preferences in Nine Wilderness and Other Roadless Areas.* USDA Forest Service Research Paper INT-253.

Lucas, R. 1981. *Redistributing Wilderness Use Through Information Supplied to Visitors.* USDA Forest Service Research Paper INT-277.

Lucas, R. 1982. Recreation Regulations—When Are They Needed? *Journal of Forestry* 80:148–51.

Lucas, R. 1983. The Role of Regulations in Recreation Management. *Western Wildlands* 9:6–10.

Lucas, R. 1985. Recreation Trends and Management of the Bob Marshall Wilderness Complex. *Proceedings of the 1985 National Outdoor Recreation Trends Symposium, Volume II.* Atlanta: U.S. National Park Service, 309–16.

Lucas, R., and G. Stankey. 1974. Social Carrying Capacity for Backcountry Recreation.

Outdoor Recreation Research: Applying the Results. USDA Forest Service General Technical Report NC-9, 14–23.

Lundgren, A., ed. 1996. *Recreation Fees in the National Park Service—Issues, Policies and Guidelines for Future Action*. St. Paul: University of Minnesota Cooperative Park Studies Unit.

Maas, A., and R. Anderson. 1986. *. . . and the Desert Shall Rejoice: Conflict, Growth and Justice in Arid Environments*. Malabur, FL: R. E. Krieger.

MacCrimmon, K., and M. Toda. 1969. The Experimental Determination of Indifference Curves. *Review of Economic Studies* 36:433–51.

Mackenzie, J. 1993. A Comparison of Contingent Preference Models. *American Journal of Agricultural Economics* 75:593–603.

Magill, A. 1976. Campsite Reservation Systems: The Campers' Viewpoint. USDA Forest Service Research Paper PSW-121.

Malthus, T. 2003 [1798]. *An Essay on the Principle of Population*. New York: W. W. Norton & Co.

Manfredo, M. 1989. An Investigation of the Basis for External Information Search in Recreation and Tourism. *Leisure Sciences* 11:29–45.

Manfredo, M., ed. 1992. *Influencing Human Behavior: Theory and Applications in Recreation, Tourism, and Natural Resources Management*. Champaign, IL: Sagamore Publishing.

Manfredo, M., and A. Bright. 1991. A Model for Assessing the Effects of Communication on Recreationists. *Journal of Leisure Research* 23:1–20.

Manfredo, M., S. Yuan, and F. McGuire. 1992. The Influence of Attitude Accessibility on Attitude-Behavior Relationships: Implications for Recreation Research. *Journal of Leisure Research* 24:157–70.

Manning, R. 1979. Strategies for Managing Recreational Use of National Parks. *Parks* 4:13–15.

Manning, R. 1986. Crowding Norms in Backcountry Settings: A Review and Synthesis. *Journal of Leisure Research* 17:75–89.

Manning, R. 1987. *The Law of Nature: Park Rangers in Yosemite Valley*. Brookline, MA: Umbrella Films.

Manning, R. 1997. Social Carrying Capacity of Parks and Outdoor Recreation Areas. *Parks and Recreation* 32:32–38.

Manning, R. 1999. *Studies in Outdoor Recreation: Search and Research for Satisfaction*. 2nd ed. Corvallis: Oregon State University Press.

Manning, R. 2001. Visitor Experience and Resource Protection: A Framework for Managing the Carrying Capacity of National Parks. *Journal of Park and Recreation Administration* 19:93–108.

Manning, R. 2003. Emerging Principles for Using Information/Education in Wilderness Management. *International Journal of Wilderness* 9:20–27, 12.

Manning, R. 2004. Recreation Planning Frameworks. *Society and Natural Resources: A Summary of Knowledge*. Jefferson, MO: Modern Litho, 83–96.

Manning, R. 2005. The Limits of Tourism in Parks and Protected Areas: Managing Carrying Capacity in the U.S. National Parks. *Taking Tourism to the Limits: Concepts, Management, Practice*. New York: Pergamon Press.

Manning, R., and S. Baker. 1981. Discrimination Through User Fees: Fact or Fiction? *Parks and Recreation* 16:70–74.

Manning, R., N. Ballinger, J. Marion, and J. Roggenbuck. 1996c. Recreation Management in Natural Areas: Problems and Practices, Status and Trends. *Natural Areas Journal* 16:142–46.

Manning, R., and M. Budruk. 2004. *Research to Support Carrying Capacity Analysis at Boston Harbor Islands National Park Area.* Burlington: University of Vermont, Park Studies Lab.

Manning, R., M. Budruk, W. Valliere, and J. Hallo. 2005. *Research to Support Visitor Management at Muir Woods National Monument and Muir Beach: Study Completion Report.* Burlington: University of Vermont, Park Studies Lab.

Manning, R., E. Callinan, H. Echelberger, E. Koenemann, and D. McEwen. 1984. Differential Fees: Raising Revenue, Distributing Demand. *Journal of Park and Recreation Administration* 2:20–38.

Manning, R., D. Cole, M. Lee, W. Stewart, and J. Taylor. 1997b. *Day Use Hiking in Grand Canyon National Park.* Burlington: University of Vermont, Park Studies Lab.

Manning, R., and W. Freimund. 2004a. Use of Visual Research Methods to Measure Standards of Quality for Parks and Outdoor Recreation. *Journal of Leisure Research* 36(4): 552–79.

Manning, R., W. Freimund, and J. Marion. 2004b. *Research to Support Application of the Visitor Experience and Resource Protection (VERP) Framework to Backcountry Planning at Zion National Park.* Burlington: University of Vermont, Park Studies Lab.

Manning, R., and C. Ginger. Forthcoming. An Owner's Manual to "Ownership": A Reply to LaChapelle and McCool. *Society and Natural Resources.*

Manning, R., A. Graefe, and S. McCool. 1996d. Trends in Carrying Capacity Planning and Management. *Proceedings of the Fourth International Outdoor Recreation and Tourism Trends Symposium.* St. Paul: University of Minnesota, 334–41.

Manning, R., C. Jacobi, and J. Marion. 2006. Recreation Monitoring at Acadia National Park. *The George Wright Forum* 23:59–72.

Manning, R., C. Jacobi, W. Valliere, and B. Wang. 1998a. Standards of Quality in Parks and Recreation. *Parks and Recreation* 33:88–94.

Manning, R., D. Johnson, and M. Vande Kamp. 1996e. Norm Congruence Among Tour Boat Passengers to Glacier Bay National Park. *Leisure Sciences* 18:125–41.

Manning, R., and E. Koeneman. 1986. Differential Campsite Pricing: An Experiment. *Campgrounds: New Perspectives on Management.* Carbondale: Southern Illinois University, 39–48.

Manning, R., W. LaPage, K. Griffall, and B. Simon. 1996f. Suggested Principles for Designing and Implementing User Fees and Charges in the National Park System. *Recreation Fees in the National Park System.* St. Paul: University of Minnesota Cooperative Park Studies Unit, 134–36.

Manning, R., and S. Lawson. 2002. Carrying Capacity as "Informed Judgment": The Values of Science and the Science of Values. *Environmental Management* 30:157–68.

Manning, R., S. Lawson, and L. Frymier. 1999b. Navigating the Confluence of Two Streams of Social Research: Contingent Valuation and Normative Standards. *Human Ecology Review* 6:35–48.

Manning, R., S. Lawson, P. Newman, M. Budruk, W. Valliere, D. Laven, and J. Bacon. 2004a. Visitor Perceptions of Recreation-Related Resource Impacts. *Environmental Impacts of Ecotourism.* London: CAB International, 273–88.

Manning, R., S. Lawson, P. Newman, D. Laven, W. Valliere. 2002b. Methodological Issues in Measuring Crowding-Related Norms in Outdoor Recreation. *Leisure Sciences* 24:339–48.

Manning, R., S. Lawson, W. Valliere, J. Bacon, and D. Laven. 2002d. *Schoodic Peninsula,*

Acadia National Park: Visitor Study 2000–2001. Burlington: University of Vermont, Park Studies Lab.

Manning, R., Y. Leung, and M. Budruk. 2005a. Research to Support Management of Visitor Carrying Capacity at Boston Harbor Islands. *Northeastern Naturalist* 12:201–20.

Manning, R., and D. Lime. 1996. Crowding and Carrying Capacity in the National Park System: Toward a Social Science Research Agenda. *Crowding and Congestion in the National Park System: Guidelines for Management and Research*. St. Paul: University of Minnesota Agricultural Experiment Station Publication 86, 27–65.

Manning, R., and D. Lime. 2000. Defining and Managing Wilderness Recreation Experiences. *Proceedings of the Wilderness Science in a Time of Change Conference*. USDA Forest Service Proceedings RMRS-P-15-Vol-4, 13–52.

Manning, R., D. Lime, W. Freimund, and D. Pitt. 1996a. Crowding Norms at Frontcountry Sites: A Visual Approach to Setting Standards of Quality. *Leisure Sciences* 18:39–59.

Manning, R., D. Lime, and M. Hof. 1996b. Social Carrying Capacity of Natural Areas: Theory and Application in the U.S. National Parks. *Natural Areas Journal* 16:118–27.

Manning, R., D. Lime, M. Hof, and W. Freimund. 1995a. The Visitor Experience and Resource Protection Process: The Application of Carrying Capacity to Arches National Park. *The George Wright Forum* 12:41–55.

Manning, R., D. Lime, M. Hof, and W. Freimund. 1995c. The Carrying Capacity of National Parks: Theory and Application. *Proceedings of the Conference on Innovations and Challenges in the Management of Visitor Opportunities in Parks and Protected Areas*. Waterloo, Canada: University of Waterloo, 9–21.

Manning, R., D. Lime, and R. McMonagle. 1995b. Indicators and Standards of the Quality of the Visitor Experience at a Heavily-Used National Park. *Proceedings of the 1994 Northeastern Recreation Recreation Research Symposium*. USDA Forest Service General Technical Report NE-198, 24–32.

Manning, R., D. Lime, R. McMonagel, and P. Nordin. 1993. *Indicators and Standards of Quality for the Visitor Experience at Arches National Park: Phase 1 Research*. St. Paul: University of Minnesota Cooperative Park Studies Unit.

Manning, R., and L. Moncrief. 1979. Land Use Analysis through Matrix Modeling: Theory and Application. *Journal of Environmental Management* 9:33–40.

Manning, R., P. Newman, E. Pilcher, J. Hallo, W. Valliere, M. Savidge, and D. Dugan. 2006. Understanding and Managing Soundscapes in the National Parks: Part 2—Standards of Quality. Paper presented at the 12th International Symposium on Science and Resource Management, Vancouver, British Columbia, Canada.

Manning, R., P. Newman, W. Valliere, B. Wang, and S. Lawson. 2001. Respondent Self-Assessment of Research on Crowding Norms in Outdoor Recreation. *Journal of Leisure Research* 33:251–71.

Manning, R., and F. Potter. 1984. Computer Simulation as a Tool in Teaching Park and Wilderness Management. *Journal of Environmental Education* 15:3–9.

Manning, R., L. Powers, and C. Mock. 1982. Temporal Distribution of Forest Recreation: Problems and Potential. *Forest and River Recreation: Research Update*. St. Paul: University of Minnesota Agricultural Experiment Station Miscellaneous Publication, 18, 26–32.

Manning, R., W. Valliere, and C. Jacobi. 1997. Crowding Norms for the Carriage Roads of Acadia National Park: Alternative Measurement Approaches. *Proceedings of the 1996 Northeastern Recreation Research Symposium*. USDA Forest Service General Technical Report NE-232, 139–45.

Manning, R., W. Valliere, S. Lawson, D. Laven, and J. Bacon. 2002e. *Blue Ridge Parkway*

Visitor Survey: Study Completion Report. Burlington: University of Vermont, Park Studies Lab.

Manning, R., W. Valliere, B. Minteer, B. Wang, and C. Jacobi. 2000. Crowding in Parks and Outdoor Recreation: A Theoretical, Empirical, and Managerial Analysis. *Journal of Park and Recreation Administration* 18:57–72.

Manning, R., W. Valliere, B. Wang, and C. Jacobi. 1999a. Crowding Norms: Alternative Measurement Approaches. *Leisure Sciences* 21:97–115.

Manning, R., W. Valliere, B. Wang, S. Lawson, and P. Newman. 2003. Estimating Day Use Social Carrying Capacity in Yosemite National Park. *Leisure: The Journal of the Canadian Association for Leisure Studies* 27:77–102.

Manning, R., W. Valliere, B. Wang, S. Lawson, and J. Treadwell. 1999c. *Research to Support Visitor Management at Statue of Liberty/Ellis Island National Monuments.* Burlington: University of Vermont, Park Studies Lab.

Manning, R., and B. Wang. 2005. Acadia National Park Carriage Roads: Estimating the Effect of Increasing Use on Crowding-Related Variables. *Computer Simulation Modeling of Recreation Use: Current Status, Case Studies, and Future Directions.* USDA Forest Service General Technical Report RMRS-GTR-143, 50–54.

Manning, R., B. Wang, W. Valliere, and S. Lawson. 1998c. *Carrying Capacity Research for Yosemite Valley: Phase I Study.* Burlington: University of Vermont, Park Studies Lab.

Manning, R., B. Wang, W. Valliere, S. Lawson, and P. Newman. 2002a. Research to Estimate and Manage Carrying Capacity of a Tourist Attraction: A Study of Alcatraz Island. *Journal of Sustainable Tourism* 10:388–464.

Manning, R., and R. Zwick. 1990. The Relationship Between Quality of Outdoor Recreation Opportunities and Support for Recreation Funding. *Proceedings of the 1989 Northeastern Recreation Research Symposium.* USDA Forest Service General Technical Report NE-145, 13–18.

Marion, J. 1991. *Developing a Natural Resource Inventory and Monitoring Program for Visitor Impacts on Recreational Sites: A Procedural Manual.* Natural Resources Report NPS/NRVT/NRR-91/06. Denver, CO: USDI National Park Service, Natural Resources Publication Office.

Marion, J., and Y. Leung. 2001. Trail Resource Impacts and an Examination of Alternative Assessment Techniques. *Journal of Park and Recreation Administration* 19:17–37.

Marion, J., J. Roggenbuck, and R. Manning. 1993. *Problems and Practices in Backcountry Recreation Management: A Survey of National Park Service Managers.* Denver, CO: U.S. National Park Service Natural Resources Report NPS/NRVT/NRR-93112.

Marler, L. 1971. A Study of Anti-Litter Messages. *Journal of Environmental Education* 3:52–53.

Martin, B. 1986. Hiker's Opinions about Fees for Backcountry Recreation. *Proceedings— National Wilderness Research Conference: Current Research.* USDA Forest Service General Technical Report INT-212, 483–88.

Martin, S., S. McCool, and R. Lucas. 1989. Wilderness Campsite Impacts. Do Managers and Visitors See Them the Same? *Environmental Management* 13:623–29.

Martinson, K., and B. Shelby. 1992. Encounter and Proximity Norms for Salmon Anglers in California and New Zealand. *North American Journal of Fisheries Management* 12:559–67.

McAvoy, L. 1990. Rescue-Free Wilderness Areas. *Adventure Education.* State College, PA: Venture Publishing, 329–34.

McAvoy, L., and D. Dustin. 1981. The Right to Risk in Wilderness. *Journal of Forestry* 79:150–52.

McAvoy, L., and D. Dustin. 1983. Indirect Versus Direct Regulation of Recreation Behavior. *Journal of Park and Recreation Administration* 1:12–17.

McAvoy, L., D. Dustin, J. Rankin, and A. Frakt. 1985. Wilderness and Legal-Liability: Guidelines for Resource Managers and Program Leaders. *Journal of Park and Recreation Administration* 3:41–49.

McCarville, R. 1996. The Importance of Price Last Paid in Developing Price Expectations for a Public Leisure Service. *Journal of Park and Recreation Administration* 14:52–64.

McCarville, R., and J. Crompton. 1987. Propositions Addressing Perception of Reference Price for Public Recreation Services. *Leisure Sciences* 9:281–91.

McCarville, R., S. Reiling, and C. White. 1986. The Role of Fairness in Users' Assessments of First-Time Fees for a Public Recreation Service. *Leisure Sciences* 18:61–76.

McCool, S., R. Benson, and J. Ashor. 1986. How the Public Perceives the Visual Effects of Timber Harvesting: An Evaluation of Interest Group Preferences. *Environmental Management* 10:385–91.

McCool, S., and N. Christensen. 1996. Alleviating Congestion in Parks and Recreation Areas through Direct Management of Visitor Behavior. *Crowding and Congestion in the National Park System: Guidelines for Management and Research.* St. Paul: University of Minnesota Agriculture Experiment Station Publication 86-1996, 67–83.

McCool, S., and D. Lime. 1989. Attitudes of Visitors towards Outdoor Recreation Management Policy. *Outdoor Recreation Benchmark 1988: Proceedings of the National Outdoor Recreation Forum.* USDA Forest Service General Technical Report SE-52, 401–11.

McCool, S., D. Lime, and D. Anderson. 1977. Simulation Modeling as a Tool for Managing River Recreation. *River Recreation Management and Research Symposium Proceedings.* USDA Forest Service General Technical Report NC-28, 202–09.

McCool, S., and J. Utter. 1981. Preferences for Allocating River Recreation Use. *Water Resources Bulletin* 17:431–37.

McCool, S., and J. Utter. 1982. Recreation Use Lotteries: Outcomes and Preferences. *Journal of Forestry* 80:10–11, 29.

McDonald, C., F. Noe, and W. Hammitt. 1987. Expectations and Recreation Fees: A Dilemma for Recreation Resource Administrators. *Journal of Park and Recreation Administration* 5:1–9.

McEwen, D., and S. Tocher. 1976. Zone Management: Key to Controlling Recreational Impact in Developed Campsites. *Journal of Forestry* 74:90–91.

McKean, M. 1982. The Japanese Experience with Scarcity: Management of Traditional Common Lands. *Environmental Review* 6:63–88.

McKenzie, D., D. Hyatt, and V. McDonald. 1992. *Ecological Indicators.* Vols. 1 and 2. New York: Elsevier Applied Science.

McLean, D., and R. Johnson. 1997. Techniques for Rationing Public Recreation Services. *Journal of Park and Recreation Administration* 15:76–92.

McLeod, S. 1997. Is the Concept of Carrying Capacity Useful in Variable Environments? *Oikos* 79:529–42.

Meadows, D., J. Randers, and W. Behrens. 1972. *The Limits to Growth.* New York: Universe Books.

Mengak, K., F. Dottavio, and J. O'Leary. 1986. The Use of Importance-Performance Analysis to Evaluate a Visitor Center. *Journal of Interpretation* 11:1–13.

Merligliano, L. 1990. Indicators to Monitor the Wilderness Recreation Experience. *Managing America's Enduring Wilderness Resource*. St. Paul: University of Minnesota, 156–62.

Meyersohn, R. 1969. The Sociology of Leisure in the United States: Introduction and Bibliography, 1945–1965. *Journal of Leisure Research* 1:53–68.

Miles, M., and M. Huberman. 1994. *Qualitative Data Analysis: An Expanded Sourcebook*. Thousand Oaks, CA: Sage Publications.

Minteer, B., and R. Manning. 2003. *Reconstructing Conservation: Finding Common Ground*. Washington, DC: Island Press.

Monte-Luna, P., B. Brook, M. Zetina-Rejon, and V. Cruz-Escalona. 2004. The Carrying Capacity of Ecosystems. *Global Ecology and Biogeography* 13:485–95.

Monz, C., C. Henderson, and R. Brame. 1994. Perspectives on the Integration of Wilderness Research, Education and Management. *Wilderness—The Spirit Lives: 6th National Wilderness Conference Proceedings*. National Park Services, Bandolier National Monument, Los Alamos, NM, 204–07.

Morehead, J. 1979. The Ranger Image. *Trends* 16:5–8.

Morrison, M. 1986. Bird Populations as Indicators of Environmental Change. *Current Ornithology*. New York: Plenum.

Moscardo, G. 1999. *Making Visitors Mindful: Volume 2*. Champaign, Illinois: Sagamore Publishing.

Muhsam, H. 1977. An Algebraic Theory of the Commons. *Managing the Commons*. San Francisco: W.H. Freeman & Co.

Muth, R., and R. Clark. 1978. *Public Participation in Wilderness and Backcountry Litter Control: A Review of Research and Management Experience*. USDA Forest Service General Technical Report PNW-75.

Nassauer, J. 1990. Using Image Capture Technology to Generate Wilderness Management Solutions. *Managing America's Enduring Wilderness Resource*. St. Paul: University of Minnesota, 553–62.

National Park Service. 1995. *VERP Implementation for Arches National Park*. Denver, CO: U.S. National Park Service.

National Park Service. 1997. *VERP: The Visitor Experience and Resource Protection (VERP) Framework—A Handbook for Planners and Managers*. Denver, CO: Denver Service Center.

National Research Council. 1986. *Proceedings of the Conference on Common Property Resource Management*. Washington, DC: National Academy Press.

National Research Council. 2000. *Ecological Indicators for the Nation*. Washington, DC: National Academy Press.

National Research Council. 2002. *The Drama of the Commons*. Washington, DC: National Academy Press.

Netting, R. 1972. Of Men and Meadows: Strategies of Alpine Land Use. *Anthropological Quarterly* 45:132–44.

Netting, R. 1976. What Alpine Peasants Have in Common: Observation on Communal Tenure in a Swiss Village. *Human Ecology* 4:135–46.

Newman, P., R. Manning, and D. Dennis. 2005. Informing Carry Capacity Decision Making in Yosemite National Park, USA, Using Stated Choice Modeling. *Journal of Park and Recreation Administration* 23:75–89.

Nicholson, W. 1995. *Microeconomic Theory: Basic Principles and Extensions*. 6th ed. Forth Worth: Dryden Press.

Nielson, C., and T. Buchanan. 1986. A Comparison of the Effectiveness of Two Interpretive Programs Regarding Fire Ecology and Fire Management. *Journal of Interpretation* 1:1–10.

Niemi, G., and M. McDonald. 2004. Application of Ecological Indicators. *Annual Review of Ecology, Evolution, and Systematics* 35:89–111.

Nunnally, J. 1978. *Psychometric Theory*. New York: McGraw-Hill.

Odum, E. 1953. *Fundamentals of Ecology*. Philadelphia: W. B. Saunders.

Odum, H. 1936. *Southern Regions of the United States*. Chapel Hill: University of North Carolina Press.

Oliver, S., J. Roggenbuck, and A. Watson. 1985. Education to Reduce Impacts in Forest Campgrounds. *Journal of Forestry* 83:234–36.

Olson, E., M. Bowman, and R. Roth. 1984. Interpretation and Nonformal Education in Natural Resources Management. *Journal of Environmental Education* 15:6–10.

Opaluch, J., S. Swallow, T. Weaver, C. Wessells, and D. Wichelns. 1993. Evaluating Impacts from Noxious Facilities: Including Public Preferences in Current Siting Mechanisms. *Journal of Environmental Economics and Management* 24:41–59.

Ostrom, E. 1990. *Governing the Commons*. Cambridge, UK: Cambridge University Press.

Ostrom, E., J. Burger, C. Field, R. Norgaard, and D. Policansky. 1999. Revisiting the Commons: Local Lessons, Global Challenges. *Science* 284:278–82.

Ostrom, V., and E. Ostrom. 1977. A Theory for Institutional Analysis of Common Pool Problems. *Managing the Commons*. San Francisco: W. H. Freeman & Co., 157–72.

Outdoor Recreation Resources Review Commission. 1962. *Outdoor Recreation for America*. Washington, DC: U.S. Government Printing Office.

Park Studies Lab. 2003. *Hawaii Volcanoes National Park: Visitor Study Completion Report*. Burlington: University of Vermont, Park Studies Lab.

Park Studies Lab. 2004. *Haleakala National Park Visitor Study: Study Completion Report*. Burlington: University of Vermont, Park Studies Lab.

Parsons, D., T. Stohlgren, and P. Fodor. 1981. Establishing Backcountry Use Quotas: The Example from Mineral King, California. *Environmental Management* 5:335–40.

Parsons, D., T. Stohlgren, and J. Kraushaar. 1982. Wilderness Permit Accuracy: Differences Between Reported and Actual Use. *Environmental Management* 6:329–35.

Patterson, M., and W. Hammitt. 1990. Backcountry Encounter Norms, Actual Reported Encounters, and Their Relationship to Wilderness Solitude. *Journal of Leisure Research* 22:259–75.

Patton, M. 2002. *Qualitative Research and Evaluation Methods*. Thousand Oaks, CA: Sage Publications.

Pearl, R., and L. Reed. 1920. On the Rate of Growth of the Population of the United States Since 1790 and Its Mathematical Representation. *Proceedings of the National Academy of Sciences* 6:275–88.

Perry, M. 1983. Controlling Crime in the Parks. *Parks and Recreation* 18: 49–51, 67.

Peterson, D. 1987. Look Ma, No Hands! Here's What's Wrong with No-Rescue Wilderness. *Parks and Recreation* 22:39–43, 54.

Peterson, G. 1992. Using Fees to Manage Congestion at Recreation Areas. *Park Visitor Research for Better Management: Park Visitor Research Workshop*. Canberra, Australia: Phillip Institute of Technology, 57–67.

Peterson, G., and D. Lime. 1979. People and Their Behavior: A Challenge for Recreation Management. *Journal of Forestry* 77:343–46.

Philley, M., and S. McCool. 1981. Law Enforcement in the National Park System: Perceptions and Practices. *Leisure Sciences* 4:355–71.

Pidd, M. 1992. *Computer Simulation in Management Science*. New York: John Wiley & Sons.

Pierskalla, C., D. Anderson, and D. Lime. 1996. Isle Royale National Park 1996 Visitor Survey: Final Report. Cooperative Park Studies Unit unpublished report. St. Paul: University of Minnesota.

Pierskalla, C., D. Anderson, and D. Lime. 1997. Isle Royale National Park 1997 Visitor Survey: Final Report. Cooperative Park Studies Unit unpublished report. St. Paul: University of Minnesota.

Pilcher, E., P. Newman, and R. Manning. 2006. Understanding and Managing Soundscapes in the National Parks: Part 1—Indicators of Quality. Paper presented at the 12th International Symposium on Society and Resource Management, Vancouver, British Columbia, Canada.

Pindyck, R., and D. Rubinfeld. 1995. *Microeconomics*. Englewood Cliffs, NJ: Prentice Hall.

Pitt, D. 1990. Developing an Image Capture System to See Wilderness Management Solutions. *Managing America's Enduring Wilderness Resource*. St. Paul: University of Minnesota, 541–48.

Pitt, D., J. Nassauer, D. Lime, and D. Snyder. 1993. The Validity of Video Imaging Presentation Media as Compared with Photographic Slides. Unpublished manuscript, University of Minnesota, St Paul.

Plager, A., and P. Womble. 1981. Compliance with Backcountry Permits in Mount McKinley National Park. *Journal of Forestry* 79:155–56.

Potter, F., and R. Manning. 1984. Application of the Wilderness Travel Simulation Model to the Appalachian Trail in Vermont. *Environmental Management* 8: 543–50.

Powers, R., J. Osborne, and E. Anderson. 1973. Positive Reinforcement of Litter Removal in the Natural Environment. *Journal of Applied Behavioral Analysis* 6:579–80.

Price, D. 1999. Carrying Capacity Reconsidered. *Population and Environment: A Journal of Interdisciplinary Studies* 21:5–26.

Ramthun, R. 1996. Information Sources and Attitudes of Mountain Bikers. *Proceedings of the 1995 Northeastern Recreation Research Symposium*. USDA Forest Service General Technical Report NE-218, 14–16.

Read, D., and S. LeBlanc. 2003. Population Growth, Carrying Capacity, and Conflict. *Current Anthropology* 44:59–85.

Rechisky, A., and B. Williamson. 1992. Impact of User Fees on Day Use Attendance at New Hampshire State Parks. *Proceedings of the 1991 Northeastern Recreation Research Symposium*. USDA Forest Service General Technical Report NE-160, 106–8.

Reiling, S., and H. Cheng. 1994. Potential Revenues from a New Day-Use Fee. *Proceedings of the 1994 Northeastern Recreation Research Symposium*. USDA Forest Service General Technical Report NE-198, 57–60.

Reiling, S., H. Cheng, C. Robinson, R. McCarville, and C. White. 1996. Potential Equity Effects of a New Day-Use Fee. *Proceedings of the 1995 Northeastern Recreation Research Symposium*. USDA Forest Service General Technical Report NE-218, 27–31.

Reiling, S., H. Cheng, and C. Trott. 1992. Measuring the Discriminatory Impact Associated with Higher Recreational Fees. *Leisure Sciences* 14:121–37.

Reiling, S., G. Criner, and S. Oltmanns. 1988. The Influence of Information on Users' Attitudes Toward Campground User Fees. *Journal of Leisure Research* 20:208–17.

Reiling, S., and M. Kotchen. 1996. Lessons Learned from Past Research on Recreation Fees. *Recreation Fees in the National Park Service: Issues, Policies and Guidelines for Future Action*. St. Paul: University of Minnesota Cooperative Park Studies Unit, 49–69.

Ribe, R. 1989. The Aesthetics of Forestry: What has Empirical Preference Research Taught Us? *Environmental Management* 13:55–74.

Richer, J., and N. Christensen. 1999. Appropriate Fees for Wilderness Day Use: Pricing Decisions for Recreation on Public Land. *Journal of Leisure Research* 31:269–80.

Robertson, R. 1982. *Visitor Knowledge Affects Visitor Behavior. Forest and River Recreation: Research Update.* St. Paul: University of Minnesota Agricultural Experiment Station Miscellaneous Publication 18, 49–51.

Roggenbuck, J. 1992. *Use of Persuasion to Reduce Resource Impacts and Visitor Conflicts. Influencing Human Behavior: Theory and Applications in Recreation, Tourism, and Natural Resources.* Champaign, IL: Sagamore Publishing, 149–208.

Roggenbuck, J., and D. Berrier. 1981. Communications to Disperse Wilderness Campers. *Journal of Forestry* 75:295–97.

Roggenbuck, J., and D. Berrier. 1982. A Comparison of the Effectiveness of Two Communication Strategies in Dispersing Wilderness Campers. *Journal of Leisure Research* 14:77–89.

Roggenbuck, J., and S. Ham. 1986. Use of Information and Education in Recreation Management. *A Literature Review: The President's Commission on Americans Outdoors.* Washington, DC: U.S. Government Printing Office, M-59–M-71.

Roggenbuck, J., J. Marion, and R. Manning. 1994. Day Users of the Backcountry: The Neglected National Park Visitor. *Trends.* 31:19–24.

Roggenbuck, J., and J. Passineau. 1986. Use of the Field Experiment to Assess the Effectiveness of Interpretation. *Proceedings of the Southeastern Recreation Research Conference.* Athens: University of Georgia Institute of Community and Area Development, 65–86.

Roggenbuck, J., and R. Schreyer. 1977. Relations Between River Trip Motives and Perception of Crowding, Management Preference, and Experience Satisfaction. *Proceedings: River Recreation Management and Research Symposium.* USDA Forest Service General Technical Report NC-28, 359–64.

Roggenbuck, J., D. Williams, S. Bange, and D. Dean. 1991. River Float Trip Encounter Norms: Questioning the Use of the Social Norms Concept. *Journal of Leisure Research* 23:133–53.

Roggenbuck, J., D. Williams, and C. Bobinski. 1992. Public-Private Partnership to Increase Commercial Tour Guides' Effectiveness as Nature Interpreters. *Journal of Park and Recreation Administration* 10:41–50.

Roggenbuck, J., D. Williams, and A. Watson. 1993. Defining Acceptable Conditions in Wilderness. *Environmental Management* 17:187–97.

Rohrmann, B., and I. Bishop. 2002. Subjective Responses to Computer Simulations of Urban Environments. *Journal of Environmental Psychology* 22:319–30.

Rosenthal, D., J. Loomis, and G. Peterson. 1984. Pricing for Efficiency and Revenue in Public Recreation Areas. *Journal of Leisure Research* 16:195–208.

Ross, T., and G. Moeller. 1974. *Communicating Rules in Recreation Areas.* USDA Forest Service Research Paper NE-297.

Rowe, R., R. d'Arge, and D. Brookshire. 1980. An Experiment on the Economic Value of Visibility. *Journal of Environmental Economics and Management* 7:1–19.

Ruddell, E., and J. Gramann. 1994. Goal Orientation, Norms, and Noise Induced Conflict among Recreation Area Users. *Leisure Sciences* 16:93–104.

Runte, A. 1990. *Yosemite: The Embattled Wilderness.* Lincoln: University of Nebraska Press.

Runte, A. 1997. *National Parks: The American Experience*. Lincoln: University of Nebraska Press.

Schechter, M., and R. Lucas. 1978. *Simulation of Recreation Use for Park and Wilderness Management*. Baltimore: Johns Hopkins University Press.

Schkade, D., and J. Payne. 1994. How People Respond to Contingent Valuation Questions: A Verbal Protocol Analysis of Willingness to Pay for Environmental Regulation. *Journal of Environmental Economics and Management* 26:88–109.

Schneider, I., and M. Budruk. 1999. Displacement as a Response to the Federal Recreation Fee Program. *Journal of Park and Recreation Administration* 17:76–84.

Schomaker, J. 1984. Writing Quantifiable River Recreation Management Objectives. *Proceedings of the 1984 National River Recreation Symposium*, 249–53.

Schomaker, J., and E. Leatherberry. 1983. A Test for Inequity in River Recreation Reservation Systems. *Journal of Soil and Water Conservation* 38:52–56.

Schroeder, H., and J. Louviere. 1999. Stated Choice Models for Predicting the Impact of User Fees at Public Recreation Sites. *Journal of Leisure Research* 31:300–24.

Schuett, M. 1993. Information Sources and Risk Recreation: The Case of Whitewater Kayakers. *Journal of Park and Recreation Administration* 11:67–72.

Schultz, J., L. McAvoy, and D. Dustin. 1988. What Are We in Business For? *Parks and Recreation* 23:52–53.

Schwartz, E. 1973. Police Services in the Parks. *Parks and Recreation* 8:72–74.

Scott, A. 1955. The Fishery: The Objectives of Sole Ownership. *Journal of Political Economy* 63:116–24.

Scott, D., and W. Munson. 1994. Perceived Constraints to Park Usage among Individuals with Low Incomes. *Journal of Park and Recreation Administration* 12:79–96.

Seidl, I., and C. Tisdell. 1999. Carrying Capacity Reconsidered: From Malthus' Population Theory to Cultural Carrying Capacity. *Ecological Economics* 31:395–408.

Seiden, E. 1954. On the Problem of Construction of Orthogonal Arrays. *Annals of Mathematical Statistics* 25:151–56.

Seig, G., J. Roggenbuck, and C. Bobinski. 1988. The Effectiveness of Commercial River Guides as Interpreters. *Proceedings of the 1987 Southeastern Recreation Research Conference*. Athens: University of Georgia, 12–20.

Shafer, C., and W. Hammitt. 1994. Management Conditions, and Indicators of Importance in Wilderness Recreation Experiences. *Proceedings of the 1993 Southeastern Recreation Research Conference*. USDA Forest Service General Technical Report SE-90, 57–67.

Shanks, B. 1976. Guns in the Parks. *The Progressive* 40:21–23.

Shelby, B. 1981. Encounter Norms in Backcountry Settings: Studies of Three Rivers. *Journal of Leisure Research* 13:129–38.

Shelby, B., N. Bregenzer, and R. Johnson. 1988b. Displacement and Product Shift: Empirical Evidence from Oregon Rivers. *Journal of Leisure Research* 20:274–88.

Shelby, B., T. Brown, and R. Baumgartner. 1992. Effects of Streamflows on River Trips on the Colorado River in Grand Canyon, Arizona. *Rivers* 3:191–201.

Shelby, B., and R. Colvin. 1982. Encounter Measures in Carrying Capacity Research: Actual, Reported, and Diary Contacts. *Journal of Leisure Research* 14:350–60.

Shelby, B., B. Danley, M. Gibbs, and M. Peterson. 1982. Preferences of Backpackers and River Runners for Allocation Techniques. *Journal of Forestry* 80:416–19.

Shelby, B., and T. Heberlein. 1984. A Conceptual Framework for Carrying Capacity Determination. *Leisure Sciences* 6:433–51.

Shelby, B., and T. Heberlein. 1986. *Carrying Capacity in Recreation Settings.* Corvallis: Oregon State University Press.

Shelby, B., and B. Shindler. 1992. Interest Group Standards for Ecological Impacts at Wilderness Campsites. *Leisure Sciences* 14:17–27.

Shelby, B., and J. Vaske. 1991. Using Normative Data to Develop Evaluative Standards for Resource Management: A Comment on Three Recent Papers. *Journal of Leisure Research* 23:173–87.

Shelby, B., J. Vaske, and M. Donnelly. 1996. Norms, Standards and Natural Resources. *Leisure Sciences* 18:103–23.

Shelby, B., J. Vaske, and R. Harris. 1988a. User Standards for Ecological Impacts at Wilderness Campsites. *Journal of Leisure Research* 20:245–56.

Shelby, B., and D. Whittaker. 1995. Flows and Recreation Quality on the Dolores River: Integrating Overall and Specific Evaluations. *Rivers* 5:121–32.

Shelby, B., D. Whittaker, and M. Danley. 1989. Allocation Currencies and Perceived Ability to Obtain Permits. *Leisure Sciences* 11:137–44.

Shindler, B. 1992. Countering the Law of Diminishing Standards. *Defining Wilderness Quality: The Role of Standards in Wilderness Management—A Workshop Proceedings.* USDA Forest Service General Technical Report PNW-305, 53–60.

Shindler, B., and B. Shelby. 1992. User Assessment of Ecological and Social Campsite Attributes. *Defining Wilderness Quality: The Role of Standards in Wilderness Management —A Workshop Proceedings.* USDA Forest Service General Technical Report PNW-305, 107–14.

Shindler, B., and B. Shelby. 1993. Regulating Wilderness Use: An Investigation on User Group Support. *Journal of Forestry* 19:41–44.

Shindler, B., and B. Shelby. 1995. Product Shift in Recreation Settings: Findings and Implications from Panel Research. *Leisure Sciences* 17:91–104.

Shuttleworth, S. 1980. The Use of Photographs as an Environmental Presentation Medium in Landscape Studies. *Journal of Environmental Management* 11:61–76.

Smith, A. 1776 (1999). *The Wealth of Nations.* London: Penguin Books.

Smith, K., and R. Headly. 1975. The Use of Computer Simulation Models in Wilderness Management. *Management Science Applications to Leisure Time.* Amsterdam: North Holland.

Smith, K., and J. Krutilla. 1976. *Structure and Properties of a Wilderness Travel Simulator.* Baltimore: Johns Hopkins University Press.

Smyth, R., M. Watzin, and R. Manning. Forthcoming. Defining Acceptable Levels for Ecosystem Indicators: Integrating Ecological Understanding and Social Values. *Environmental Management.*

Society of American Foresters. 1993. *Sustaining Long-Term Forest Health and Productivity.* Bethesda: Society of American Foresters.

Stamps, A. 1990. Use of Photographs to Simulate Environments: A Meta-analysis. *Perceptual and Motor Skills* 71:907–13.

Stanfield, R., R. Manning, M. Budruk, and M. Floyd. 2006. Racial Discrimination in Parks and Outdoor Recreation: An Empirical Study. *Proceedings of the 2005 Northeastern Recreation Research Symposium.* USDA Forest Service General Technical Report NE-341, 247–53.

Stankey, G. 1973. *Visitor Perception of Wilderness Recreation Carrying Capacity.* USDA Forest Service Research Paper INT-142.

Stankey, G. 1975. Criteria for the Determination of Recreational Carrying Capacity in the Colorado River Basin. *Environmental Management in the Colorado River Basin.* Logan: Utah State University Press.

Stankey, G. 1979. Use Rationing in Two Southern California Wildernesses. *Journal of Forestry* 77:347–49.

Stankey, G. 1980a. *A Comparison of Carrying Capacity Perceptions Among Visitors to Two Wildernesses.* USDA Forest Service Research Paper INT-242.

Stankey, G. 1980b. Wilderness Carrying Capacity: Management and Research Progress in the United States. *Landscape Research* 5:6–11.

Stankey, G. 1989. Solitude for the Multitudes: Managing Recreational Use in Wilderness. *Public Places and Spaces.* New York: Plenum Press, 277–99.

Stankey, G., and J. Baden. 1977. *Rationing Wilderness Use: Methods, Problems, and Guidelines.* USDA Forest Service Research Paper INT-192.

Stankey, G., R. Clark, and B. Bormann. 2005. *Adaptive Management of Natural Resources: Theory, Concepts, and Management Institutions.* USDA Forest Service General Technical Report PNW-GTR-654.

Stankey, G., D. Cole, R. Lucas, M. Peterson, S. Frissell, and R. Washburne. 1985. *The Limits of Acceptable Change (LAC) System for Wilderness Planning.* USDA Forest Service General Technical Report INT-176.

Stankey, G., and D. Lime. 1973. *Recreational Carrying Capacity: An Annotated Bibliography.* USDA Forest Service General Technical Report INT-3.

Stankey, G., and R. Manning. 1986. Carrying Capacity of Recreation Settings. *A Literature Review: The President's Commission on Americans Outdoors.* Washington, DC: U.S. Government Printing Office, M-47–M-57.

Stankey, G., S. McCool, and G. Stokes. 1984. Limits of Acceptable Change: A New Framework for Managing the Bob Marshall Wilderness Complex. *Western Wildlands* 10:33–37.

Stankey, G., and R. Schreyer. 1987. Attitudes toward Wilderness and Factors Affecting Visitor Behavior: A State-of-Knowledge Review. *Proceedings—National Wilderness Research Conference: Issues, State-of-Knowledge, Future Directions.* USDA Forest Service General Technical Report INT-220, 246–93.

Stevenson, S. 1989. A Test of Peak Load Pricing on Senior Citizen Recreationists: A Case Study of Steamboat Lake State Park. *Journal of Park and Recreation Administration* 7:58–68.

Stewart, W. 1989. Fixed Itinerary Systems in Backcountry Management. *Journal of Environmental Management* 29:163–71.

Stewart, W. 1991. Compliance with Fixed Itinerary Systems in Water-Based Parks. *Environmental Management* 15:235–40.

Stewart, W., and D. Cole. 2001. Number of Encounters and Experience Quality in Grand Canyon Backcountry: Consistently Negative and Weak Relationships. *Journal of Leisure Research* 3:106–20.

Stewart, W., and D. Cole. 2003. On the Prescriptive Utility of Visitor Survey Research: A Rejoinder to Manning. *Journal of Leisure Research.* 35:119–27.

Stewart, W., D. Cole, R. Manning, W. Valiere, J. Taylor, and M. Lee. 2000. Preparing for a Day Hike at Grand Canyon: What Information Is Useful? *Wilderness Science in a Time of Change Conference.* Ogden, UT: USDA Forest Service, Rocky Mountain Research Station. RMRS-P-15-VOL-4.

Sumner, E. 1936. *Special Report on a Wildlife Study in the High Sierra in Sequoia and*

Yosemite National Parks and Adjacent Territory. Washington, DC: U.S. National Park Service Records, National Archives.

Swearingen, T., and D. Johnson. 1995. Visitors' Responses to Uniformed Park Employees. *Journal of Park and Recreation Administration* 13:73–85.

Tarrant, M., H. Cordell, and T. Kibler. 1997. Measuring Perceived Crowding for High-Density River Recreation: The Effects of Situational Conditions and Personal Factors. *Leisure Sciences* 19:97–112

Tashakkori, A., and C. Teddlie. 1998. *Mixed Methodology: Combining Qualitative and Quantitative Approaches.* Thousand Oaks, CA: Sage Publications.

Taylor, D., and P. Winter. 1995. Environmental Values, Ethics, and Depreciative Behavior in Wildland Settings. *Proceedings of the Second Symposium on Social Aspects and Recreation Research.* USDA Forest Service General Technical Report PSW-156, 59–66.

Thayer, M. 1981. Contingent Valuation Techniques for Assessing Environmental Impacts: Further Evidence. *Journal of Environmental Economics and Management* 8:27–44.

Titre, J., and A. Mills. 1982. Effect of Encounters on Perceived Crowding and Satisfaction. *Forest and River Recreation: Research Update* University of Minnesota Agricultural Experiment Station Miscellaneous Publication 18, 146–53.

Twight, B., K. Smith, and G. H. Wassinger. 1981. Privacy and Camping: Closeness to the Self Versus Closeness to Others. *Leisure Sciences* 4:427–41.

Underhill, A., A. Xaba, and R. Borkan. 1986. The Wilderness Use Simulation Model Applied to Colorado River Boating in Grand Canyon National Park, USA. *Environmental Management* 10: 367–74.

U.S. Environmental Protection Agency. 2002. *A SAB Report: A Framework for Assessing and Reporting on Ecological Conditions.* Washington, DC: U.S. Environmental Protection Agency.

Utter, J., W. Gleason, and S. McCool. 1981. User Perceptions of River Recreation Allocation Techniques. *Some Recent Products of River Recreation Research.* USDA Forest Service General Technical Report NC-63, 27–32.

Uysal, M., C. McDonald, and L. Reid. 1990. Sources of Information Used by International Visitors to U.S. Parks and Natural Areas. *Journal of Park and Recreation Administration* 8:51–59.

Vande Kamp, M., J. Swanson, and D.R. Johnson. 2003. *Social Science Research for Managing the Exit Glacier Fee Area of Kenai Fjords National Park: Volume 1, Visitor Experiences and Visitor Use Levels.* Technical Report NPS/CCSOUW/NRTR-2003-05. PNW Cooperative Ecosystem Studies Unit, National Park Service, College of Forest Resources, University of Washington.

Valliere, W., and R. Manning. 2003. Applying the Visitor Experience and Resource Protection (VERP) Framework to Cultural Resources in the National Parks. *Proceedings of the 2002 Northeastern Recreation Research Symposium.* USDA Forest Service General Technical Report NE-302, 234–38.

Valliere, W., L. Park, J. Hallo, R. McCown, and R. Manning. 2006. Enhancing Visual Research with Computer Animation: A Study of Crowding-Related Standards of Quality for the Loop Road at Acadia National Park. Paper presented at the 2006 12th International Symposium on Society and Resource Management, Vancouver, British Columbia, Canada.

Van Wagtendonk, J. 1981. The Effect of Use Limits on Backcountry Visitation Trends in Yosemite National Park. *Leisure Sciences* 4:311–23.

Van Wagtendonk, J., and J. Benedict. 1980. Wilderness Permit Compliance and Validity. *Journal of Forestry* 78:399–401.

Van Wagtendonk, J., and P. Coho. 1986. Trailhead Quotas: Rationing Use to Keep Wilderness Wild. *Journal of Forestry* 84:22–24.

Van Wagtendonk, J., and D. Cole. 2005. Historical Development of Simulation Models of Recreation Use. *Computer Simulation Modeling of Recreation Use Current Status, Case Studies, and Future Directions.* USDA Forest Service General Technical Report RMRS-GTR-143, 3–10.

Vander Stoep, G., and J. Gramann. 1987. The Effect of Verbal Appeals and Incentives on Depreciative Behavior Among Youthful Park Visitors. *Journal of Leisure Research* 19:69–83.

Vander Stoep, G., and J. Roggenbuck. 1996. Is Your Park Being "Loved to Death"?: Using Communication and Other Indirect Techniques to Battle the Park "Love Bug." *Crowding and Congestion in the National Park System: Guidelines for Research and Management.* St. Paul: University of Minnesota Agricultural Experiment Station Publication 86-1996, 85–132.

Vaske, J., M. Donnelly, and R. Deblinger. 1990. Norm Activation and the Acceptance of Behavioral Restrictions Among Over Sand Vehicle Users. *Proceedings of the 1990 Northeastern Recreation Research Symposium.* USDA Forest Service General Technical Report NE-145, 153–59.

Vaske, J., M. Donnelly, R. Doctor, and J. Petruzzi. 1995. Frontcountry Encounter Norms Among Three Cultures. *Proceedings of the 1994 Northeastern Recreation Research Symposium.* USDA Forest Service General Technical Report NE-198, 162–65.

Vaske, J., M. Donnelly, and J. Petruzzi. 1996. Country of Origin, Encounter Norms and Crowding in a Frontcountry Setting. *Leisure Sciences* 18:161–76.

Vaske, J., A. Graefe, B. Shelby, and T. Heberlein. 1986. Backcountry Encounter Norms: Theory, Method, and Empirical Evidence. *Journal of Leisure Research* 18:137–53.

Vaske, J., and D. Whittaker. 2004. Normative Approaches to Natural Resources. *Society and Natural Resources: A Summary of Knowledge.* Jefferson, MO: Modern Litho, 283–94

Vaske, J., D. Whittaker, B. Shelby, and M. Manfredo. 2000. Indicators and Standards: Developing Definitions of Quality. *Wildlife Viewing: A Management Handbook.* Corvallis: Oregon State University Press, 143–71.

Verhulst, P. 1838. Notice sure la los que la Population suit dans son Accroissement. *Correspondance Mathematique et Physique* 10:113–21.

Vining, J., and B. Orland. 1989. The Video Advantage: A Comparison of Two Environmental Representation Techniques. *Journal of Environmental Management* 29:275–83.

Vogt, C., and D. Williams. 1999. Support for Wilderness Recreation Fees: The Influence of Fee Purpose and Day Versus Overnight Use. *Journal of Park and Recreation Administration* 17:85–99.

Wade, J. 1979. Law Enforcement in the Wilderness. *Trends* 16:12–15.

Wagar, J. A. 1964. The Carrying Capacity of Wild Lands for Recreation. *Forest Science Monograph 7.* Washington, DC: Society of American Foresters.

Wagar, J. A. 1968. The Place of Carrying Capacity in the Management of Recreation Lands. *Third Annual Rocky Mountain-High Plains Park and Recreation Conference Proceedings.* Fort Collins: Colorado State University.

Wagar, J. A. 1974. Recreational Carrying Capacity Reconsidered. *Journal of Forestry* 72:274–78.

Wagar, J. V. 1946. Services and Facilities for Forest Recreationists. *Journal of Forestry* 44:883–87.

Wagar, J. V. 1951. Some Major Principles in Recreation Land Use Planning. *Journal of Forestry* 49:431–35.

Wagstaff, M., and B. Wilson. 1988. The Evaluation of Litter Behavior Modification in a River Environment. *Proceedings of the 1987 Southeastern Recreation Research Conference.* Athens: University of Georgia, 21–28.

Walsh, R. 1986. *Recreation Economic Decisions.* State College, Pa: Venture Publishing.

Walters, C. 1986. *Adaptive Management of Renewable Natural Resources.* New York: Macmillan.

Wang, B., and R. Manning. 1999. Computer Simulation Modeling for Recreation Management: A Study on Carriage Road Use in Acadia National Park, Maine, USA. *Environmental Management* 23:193–203.

Wang, B., R. Manning, S. Lawson, and W. Valliere. 2001. Estimating Social Carrying Capacity through Simulation Modeling: An Application to Arches National Park, Utah. *Proceedings of the 2000 Northeastern Recreation Research Symposium.* USDA Forest Service General Technical Report NE-276, 193–200.

Warzecha, C., R. Manning, D. Lime, and W. Freimund. 2001. Diversity in Outdoor Recreation: Planning and Managing a Spectrum of Visitor Opportunities in and Among Parks. *The George Wright Forum* 18(3):99–111.

Washburne, R. 1981. Carrying Capacity Assessment and Recreational Use in the National Wilderness Preservation System. *Journal of Soil and Water Conservation* 36:162–66.

Washburne, R. 1982. Wilderness Recreational Carrying Capacity: Are Numbers Necessary? *Journal of Forestry* 80:726–28.

Washburne, R., and D. Cole. 1983. *Problems and Practices in Wilderness Management: A Survey of Managers.* USDA Forest Service Research Paper INT-304.

Watson, A. 1993. *Characteristics of Visitors Without Permits Compared to Those With Permits at the Desolation Wilderness, California.* USDA Forest Service Research Note INT-414.

Watson, A. 1995. Opportunities for Solitude in the Boundary Waters Canoe Area Wilderness. *Northern Journal of Applied Forestry* 12:12–18.

Watson, A., and M. Niccolucci. 1995. Conflicting Goals of Wilderness Management: Natural Conditions Versus Natural Experiences. *Proceedings of the Second Symposium on Social Aspects and Recreation Research.* USDA Forest Service General Technical Report PSW-156, 11–15.

Watzin, M., R. Smyth, E. Cassell, W. Hession, R. Manning, and D. Wang. 2005. *Ecological Indicators and an Environmental Scorecard for the Lake Champlain Basin Program (LCBP Technical Report 46)* Grand Isle, VT: Lake Champlain Basin Program.

West, P. 1982. Effects of User Behavior on the Perception of Crowding in Backcountry Forest Recreation. *Forest Science* 28:95–105.

Westover, T., T. Flickenger, and M. Chubb. 1980. Crime and Law Enforcement. *Parks and Recreation* 15:28–33.

Whittaker, D. 1992. Selecting Indicators: Which Impacts Matter More? *Defining Wilderness Quality: The Role of Standards in Wilderness Quality—A Workshop Proceedings.* USDA Forest Service General Technical Report PNW-305, 13–22.

Whittaker, D., and B. Shelby. 1988. Types of Norms for Recreation Impact: Extending the Social Norms Concept. *Journal of Leisure Research* 20:261–73.

Whittaker, D., and B. Shelby. 2002. Evaluating Instream Flows for Recreation: Applying the Structural Norm Approach to Biophysical Conditions. *Leisure Sciences* 24:363–74.

Whittaker, D., and B. Shelby. 1992. Developing Good Standards: Criteria, Characteristics, and Sources. *Defining Wilderness Quality: The Role of Standards in Wilderness Management—A Workshop Proceedings.* USDA Forest Service General Technical Report PNW-305, 6–12.

Whittaker, R. 1997. Capacity Norms on Bear Viewing Platforms. *Human Dimensions of Wildlife* 2:37–49.

Wicker, A., and S. Kirmeyer. 1976. What the Rangers Think. *Parks and Recreation* 11:28–30, 42.

Wicks, B. 1987. The Allocation of Recreation and Park Resources: The Courts' Intervention. *Journal of Park and Recreation Administration* 5:1–9.

Wicks, B., and J. Crompton. 1986. Citizen and Administrator Perspectives of Equity in the Delivery of Park Services. *Leisure Sciences* 8:341–65.

Wicks, B., and J. Crompton. 1987. An Analysis of the Relationships between Equity Choice Preferences, Service Type and Decision Making Groups in a U.S. City. *Journal of Leisure Research* 19:189–204.

Wicks, B., and J. Crompton. 1989. Allocation Services for Parks and Recreation: A Model for Implementing Equity Concepts in Austin, Texas. *Journal of Urban Affairs* 11:169–88.

Wicks, B., and J. Crompton. 1990. Predicting the equity preferences of park and recreation department employees and residents of Austin, Texas. *Journal of Leisure Research* 22:18–35.

Wiklee, T. 1991. Comparing Rationing Policies Used on Rivers. *Journal of Park and Recreation Administration* 9:73–80.

Williams, D., J. Roggenbuck, and S. Bange. 1991. The Effect of Norm-Encounter Compatibility on Crowding Perceptions, Experience, and Behavior in River Recreation Settings. *Journal of Leisure Research* 23:154–72.

Williams, D., J. Roggenbuck, M. Patterson, and A. Watson. 1992. The Variability of User-Based Social Impact Standards for Wilderness Management. *Forest Science* 38:738–56.

Williams, D., C. Vogt, and J. Vitterso. 1999. Structural Equation Modeling of Users' Response to Wilderness Recreation Fees. *Journal of Leisure Research* 31:245–68.

Willis, C., J. Canavan, and R. Bond. 1975. Optimal Short-Run Pricing Policies for a Public Campground. *Journal of Leisure Research* 7:108–13.

Wilson, J. 1977. A Test of the Tragedy of the Commons. *Managing the Commons.* San Francisco: W.H. Freeman & Co., 96–111.

Wilman, E. 1988. Pricing Policies for Outdoor Recreation. *Land Economics* 64:234–41.

Winter, P., L. Palucki, and R. Burkhardt. 1999. Anticipated Responses to a Fee Program: The Key Is Trust. *Journal of Leisure Research* 31:207–26.

Wittmann, K., J. Vaske, M. Manfredo, and H. Zinn. 1998. Standards for Lethal Response to Problem Urban Wildlife. *Human Dimensions of Wildlife* 3:29–48.

Womble, P. 1979. *Survey of Backcountry Users in Mount McKinley National Park, Alaska: A Report for Management.* Seattle: National Park Service Cooperative Park Studies Unit, University of Washington.

World Commission on Environment and Development. 1987. *Our Common Future.* New York: Oxford University Press.

Young, J., D. Williams, and J. Roggenbuck. 1991. The Role of Involvement in Identifying Users' Preferences for Social Standards in the Cohutta Wilderness. *Proceedings of the 1990*

Southeastern Recreation Research Conference. USDA Forest Service General Technical Report SE-67, 173–83.

Zinn, H., M. Manfredo, and J. Vaske. 2000. Social Psychological Bases for Stakeholder Acceptance Capacity. *Human Dimensions of Wildlife* 5:20–33.

Zinn, H., M. Manfredo, J. Vaske, and K. Wittmann. 1998. Using Normative Beliefs to Determine the Acceptability of Wildlife Management Actions. *Society and National Resources* 11:649–62.

Index